Lanthanides in Organic Synthesis

BEST SYNTHETIC METHODS

Series Editors

A. R. Katritzky
University of Florida
Gainesville, Florida
USA

O. Meth-Cohn
Sunderland Polytechnic
Sunderland
UK

C.W. Rees
Imperial College of Science
and Technology
London, UK

Richard F. Heck, *Palladium Reagents in Organic Syntheses*, 1985
Alan H. Haines, *Methods for the Oxidation of Organic Compounds: Alkanes, Alkenes, Alkynes, and Arenes*, 1985
Paul N. Rylander, *Hydrogenation Methods*, 1985
Ernest W. Colvin, *Silicon Reagents in Organic Synthesis*, 1988
Andrew Pelter, Keith Smith and Herbert C. Brown, *Borane Reagents*, 1988
Basil Wakefield, *Organolithium Methods*, 1988
Alan H. Haines, *Methods for the Oxidation of Organic Compounds: Alcohols, Alcohol Derivatives, Alkyl Halides, Nitroalkanes, Alkyl Azides, Carbonyl Compounds, Hydroxyarenes and Aminoarenes*, 1988
H. G. Davies, R. H. Green, D. R. Kelly and S. M. Roberts, *Biotransformations in Preparative Organic Chemistry: The Use of Isolated Enzymes and Whole Cell Systems*, 1989
I. Ninomiya and T. Naito, *Photochemical Synthesis*, 1989
T. Shono, *Electroorganic Synthesis*, 1991
David Crich and William B. Motherwell, *Free-radical Chain Reactions in Organic Synthesis*, 1991
N. Petragnani, *Tellurium in Organic Synthesis*, 1994
T. Imamoto, *Lanthanides in Organic Synthesis*, 1994
A. J. Pearson, *Iron Compounds in Organic Synthesis*, 1994

Lanthanides in Organic Synthesis

Tsuneo Imamoto

*Department of Chemistry, Faculty of Science,
Chiba University, Inage, Chiba 263,
Japan*

ACADEMIC PRESS

Harcourt Brace & Company, Publishers
London   San Diego   New York
Boston   Sydney   Tokyo   Toronto

ACADEMIC PRESS LIMITED
24–28 Oval Road
London NW1 7DX

US Edition published by
ACADEMIC PRESS INC.
San Diego, CA 92101

This book is a guide providing general information concerning its subject matter; it is not a procedural manual. Synthesis of chemicals is a rapidly changing field. The readers should consult current procedural manuals for state-of-the-art instructions and applicable government safety regulations. The publisher and the authors do not accept responsibility for any misuse of this book, including its use as a procedural manual or as a source of specific instructions.

A CIP record for this book is available from the British Library

ISBN 0-12-370722-6

Typeset by Avant Mode Ltd, Bournemouth, Dorset
Printed and bound in Great Britain by Hartnolls Limited, Bodmin, Cornwall

Contents

Foreword

There is a vast and often bewildering array of synthetic methods and reagents available to organic chemists today. Many chemists have their own favoured methods, old and new, for standard transformations, and these can vary considerably from one laboratory to another. New and unfamiliar methods may well allow a particular synthetic step to be done more readily and in higher yield, but there is always some energy barrier associated with their use for the first time. Furthermore, the very wealth of possibilities creates an information-retrieval problem. How can we choose between all the alternatives, and what are their real advantages and limitations? Where can we find the precise experimental details, so often taken for granted by the experts? There is therefore a constant demand for books on synthetic methods, especially the more practical ones like *Organic Syntheses, Organic Reactions,* and *Reagents for Organic Synthesis,* which are found in most chemistry laboratories. We are convinced that there is a further need, still largely unfulfilled, for a uniform series of books, each dealing concisely with a particular topic from a *practical* point of view – a need, that is, for books full of preparations, practical hints and detailed examples, all critically assessed, and giving just the information needed to smooth our way painlessly into the unfamiliar territory. Such books would obviously be a great help to research students as well as to established organic chemists.

We have been very fortunate with the highly experienced and expert organic chemists, who, agreeing with our objective, have written the first group of volumes in this series, *Best Synthetic Methods*. We shall always be pleased to received comments from readers and suggestions for future volumes.

A.R.K., O.M.-C., C.W.R

Preface

Organic synthesis with lanthanides has undergone a considerable development over recent decades; a great number of synthetic reactions have been explored by the use of lanthanide reagents, and some of them have become indispensable protocols in modern organic synthesis.

The aim of this book is to describe this rapidly growing area from a practical point of view. The author has tried to summarize synthetically useful and/or novel organic transformations with emphasis on the characteristic properties of lanthanide reagents. They are concisely presented in many schemes and tables, and in some cases their synthetic applications are described briefly. Actual examples and general procedures are also given. Detailed mechanistic explanations are beyond the scope of this book, and a few only are noted at appropriate points in the text. It is hoped that the organization of the material will be useful to research students as well as to established organic chemists.

Finally, the author acknowledges an enormous debt to the editor, Professor Otto Meth-Cohn, who provided him with valuable comments and suggestions. Thanks are due to John Haynes and Roger Hill of Academic Press for giving the author the opportunity to write this book and for their linguistic improvements.

Tsuneo Imamoto
Chiba, Japan

Detailed Contents

Abbreviations

Ac	acetyl	LDA	lithium diisopropylamide
AIBN	2,2'-azobis(isobutyronitrile)	LDBB	lithium 4,4'-di-t-butylbiphenyl
Ar	aryl	LDMAN	lithium 1-dimethylaminonaphthalene
Boc	t-butoxycarbonyl	LN	lithium naphthalene
Bn	benzyl	Me	methyl
Bu	n-butyl	MMA	methyl methacrylate
Bu^i	isobutyl	MOM	methoxymethyl
Bu^t	t-butyl	NAFK	Nafion®551
CAN	cerium(IV) ammonium nitrate, ammonium cerium(IV) nitrate	NMR	nuclear magnetic resonance
		Ph	phenyl
CAS	cerium(IV) ammonium sulphate, ammonium cerium(IV) sulphate	Pr	n-propyl
		Pr^i	isopropyl
CSA	10-camphorsulphonic acid	psi	pounds per square inch
DABCO	1,4-diazabicyclo[2.2.2]octane	py	pyridine
DBM	dibenzoylmethanato	rt	room temperature
DMA	dimethylacetamide	TBDMS	t-butyldimethylsilyl
DMAE	N,N-dimethylaminoethanol	Tf	triflyl, trifluoromethanesulphonyl
DMAP	4-dimethylaminopyridine	TFA	trifluoroacetic acid
DME	1,2-dimethoxyethane	THF	tetrahydrofuran
DMF	N,N-dimethylformamide	THP	tetrahydropyran
DMPU	1,3-dimethyl-3,4,5,6-tetrahydro-2(1H)-pyrimidinone	TIPDS	3',5'-O-tetraisopropyldisoloxane-1,3-diyl
DMSO	dimethylsulphoxide	TLC	thin layer chromatography
dppm	di(perfluoro-2-propoxypropionyl)methanato	TMEDA	N,N,N',N'-tetramethyl-1,2-ethanediamine, tetramethylethylenediamine
Et	ethyl	TMS	trimethylsilyl
ether	diethyl ether	Ts	tosyl, p-toluenesulphonyl
fod	6,6,7,7,8,8,8-heptafluoro-2,2-dimethyl-3,5-octanedionato		
GC	gas chromatography		
hfc	3-(heptafluoropropylhydroxy-methylene)-(+)-camphorato		
HMPA	hexamethylphosphoric triamide, tris(dimethylamino)phosphine oxide (CAUTION – CANCER SUSPECT AGENT		

–1–

Introduction

The lanthanides are the 15 elements that belong to the third group and sixth period in the periodic table. These elements are generally called f-block elements or inner transition elements, since (with the exception of lanthanum) they possess 4f electrons. Lanthanide elements exhibit unique physical and chemical properties which differ from those of main group elements and d-block transition elements. A number of functionalized or high value-added materials such as ceramics, fluorescent materials, magnetic substances, hydrogen absorbants and so on have been produced by the use of lanthanide elements.

On the other hand, complexes of europium and praseodymium have frequently been used in the field of organic chemistry as NMR shift reagents. Tetravalent lanthanide salts such as cerium(IV) ammonium nitrate (CAN) and cerium(IV) ammonium sulphate (CAS) are also often employed in organic synthesis as oxidants. However, extensive uses of lanthanide elements were initiated only after the pioneering works by Kagan and Luche. Numerous synthetic reactions and new methodologies have been explored by utilizing the characteristic properties of lanthanides. Some of them are superior to existing methods, and are the method of choice in synthesis.

A number of review articles and monographs dealing with lanthanides in organic chemistry have been published. Representative ones are shown below.

Organic synthesis with lanthanides

1. H. B. Kagan and J. L. Namy, in *Handbook on the Physics and Chemistry of the Rare Earths*, ed. K. A. Gschneider and L. Eyring, p. 525. Elsevier, Amsterdam, 1984.
2. H. B. Kagan, in *Fundamental and Technological Aspects of Organo-f-Element Chemistry*, ed. T. J. Marks and I. L. Fragala, p. 49. Reidel, Dordrecht, 1985.
3. J. R. Long, in *Handbook on the Physics and Chemistry of the Rare Earths*, ed. K. A. Gschneider and L. Eyring, p. 335. Elsevier, Amsterdam, 1986.
4. H. B. Kagan and J. L. Namy, *Tetrahedron* **42**, 6573 (1986).
5. T.-L. Ho, in *Organic Syntheses by Oxidation with Metal Compounds*, ed. W. J. Mijs and C. R. H. I. de Jonge, p. 569. Plenum, New York, 1986.
6. G. A. Molander, in *The Chemistry of the Metal Carbon Bond*, Vol. 5, ed. F. R. Hartley, Chap. 8. John Wiley, New York, 1989.
7. T. Imamoto, in *Comprehensive Organic Synthesis*, Vol. 1, ed. B. M. Trost, I. Fleming and S. L.

Schreiber, Chap. 1.8, p. 231. Pergamon, London, 1991.
8.   G. A. Molander, in *Comprehensive Organic Synthesis,* Vol. 1, ed. B. M. Trost, I. Fleming and S. L. Schreiber, Chap. 1.9, p. 251. Pergamon, London, 1991.
9.   G. A. Molander, *Chem. Rev.* **92,** 29 (1992).

Organolanthanide complexes

1.   T. J. Marks and R. D. Ernst, in *Comprehensive Organometallic Chemistry,* Vol. 3, ed. G. Wilkinson, 1982. p. 173. Pergamon Press, Oxford.
2.   W. J. Evans, *Adv. Organomet. Chem.* **24,** 131 (1985).
3.   H. Schumann, *Angew. Chem. Int. Ed. Engl.* **23,** 474 (1984).
4.   R. D. Ernst, *J. Organomet. Chem.* **392,** 51 (1990).

Important individual lanthanide compounds

1.   *Dictionary of Organometallic Compounds,* ed. J. Buckingham, Chapman & Hall, London, 1984.
2.   J. H. Forsberg, Y. Marcus and T. Moeller, *Gmelins Handbuch der Anorganischen Chemie,* 8th Edn, Part D6. Springer-Verlag, Berlin, 1983.

$-2-$

General Properties of Lanthanides

The lanthanides are the 15 elements from lanthanum to lutetium in the periodic table. These elements constitute a large family of closely related 4f elements which sequentially occupy 4f shells. Some general data on the lanthanides are summarized in Table 2.1.

2.1   OXIDATION STATES

The most stable oxidation state of lanthanides is +3. The +2 and +4 oxidation states are unusual in lanthanide series. These oxidation states are formed by elements that can attain empty, half-filled or filled 4f shells.

For dispositive lanthanides, samarium (f^6, nearly half-filled), europium (f^7, half-filled), thulium (f^{13}, nearly filled) and ytterbium (f^{14}, filled) have been prepared and characterized. Eu^{2+} and Yb^{2+} are the most stable, though they gradually react with water. Sm^{2+} is relatively unstable and exhibits strong reducing power. Tm^{2+} is an extremely unstable species.

For tetrapositive lanthanides, Ce^{4+} (f^0), Pr^{4+} (f^1), and $Tb(f^7)$ are known. Ce^{4+} is the most stable, even in aqueous solution. Cerium(IV) salts such as cerium(IV) ammonium nitrate (CAN) and cerium(IV) ammonium sulphate (CAS) are useful oxidants for organic substrates.

The redox properties depend partly on electronic factors such as the inherent stability of the f shells. They are affected also by medium (e.g. solvent, acidity and basicity) and ligands attached to the element.

2.2   IONIC RADIUS

The ionic radii of the lanthanide ions are larger than those of the d-block transition elements. They decrease monomerically with increasing atomic number. This is the well-known "lanthanide contraction", which arises from ineffective shielding of the 4f electrons, resulting in an increase in effective nuclear charge and a concomitant decrease in ionic radius.

3

TABLE 2.1

General Properties of Lanthanide Ions

Atomic no.	Name	Symbol	Clark no. (%)	Electronic configurations				Oxidation reduction potential			Ionic radius Ln^{3+} (Å)
				Atom	M^{2+}	M^{3+}	M^{4+}	Ln/Ln^{3+}	Ln^{2+}/Ln^{3+}	Ln^{3+}/Ln^{4+}	
57	Lanthanum	La	1.8×10^{-3} (35)	$5d^16s^2$	$5d^1$	[Xe]	–	+2.52			1.172
58	Cerium	Ce	4.5×10^{-3} (28)	$4f^15d^16s^2$	$4f^2$	$4f^1$	[Xe]	2.48		−1.74	1.15
59	Praseodymium	Pr	5×10^{-4} (48)	$4f^36s^2$	$4f^3$	$4f^2$	$4f^1$	2.46		−3.2	1.13
60	Neodymium	Nd	2.2×10^{-5} (33)	$4f^46s^2$	$4f^4$	$4f^3$	$4f^2$	2.43			1.123
61	Promethium	Pm	–	$4f^56s^2$	$4f^5$	$4f^4$	–	2.42			$(1.11)^a$
62	Samarium	Sm	6×10^{-4} (44)	$4f^66s^2$	$4f^6$	$4f^5$	–	2.41	+1.55		1.098
63	Europium	Eu	1×10^{-4} (58)	$4f^76s^2$	$4f^7$	$4f^6$	–	2.41	+0.35		1.087
64	Gadolinium	Gd	6×10^{-4} (45)	$4f^75d^16s^2$	$4f^75d^1$	$4f^7$	–	2.40			1.078
65	Terbium	Tb	8×10^{-5} (59)	$4f^96s^2$	$4f^9$	$4f^8$	$4f^7$	2.39		−3.1	1.063
66	Dysprosium	Dy	4×10^{-4} (52)	$4f^{10}6s^2$	$4f^{10}$	$4f^9$	$4f^8$	2.35			1.052
67	Holmium	Ho	1×10^{-4} (57)	$4f^{11}6s^2$	$4f^{11}$	$4f^{10}$	–	2.32			1.041
68	Erbium	Er	2×10^{-4} (56)	$4f^{12}6s^2$	$4f^{12}$	$4f^{11}$	–	2.30			1.033
69	Thulium	Tm	2×10^{-5} (66)	$4f^{13}6s^2$	$4f^{13}$	$4f^{12}$	–	2.28	+2.1		1.020
70	Ytterbium	Yb	2.5×10^{-4} (55)	$4f^{14}6s^2$	$4f^{14}$	$4f^{13}$	–	2.27	+1.15		1.008
71	Lutetium	Lu	7×10^{-5} (60)	$4f^{14}5d^16s^2$	–	$4f^{14}$	–	2.25			1.001

[a]Estimated value.

Ionic radii are affected by oxidation state; the dipositive ion has ionic radii about 10% larger than those of the tripositive ion, and the ionic radius of the tetrapositive ion is smaller than that of the tripositive one.

The relatively large ionic radii of the lanthanides allow the accommodation of up to 12 ligands in the coordination sphere, and coordination numbers of 7, 8 and 9 are common. In general, "light" lanthanides, whose ionic radii are large, tend to have higher coordination numbers than "heavy" lanthanides.

2.3 OCCURRENCE

Lanthanides are often called the "rare earths". However, these elements are not really rare, with the exception of promethium which is radioactive and does not occur naturally. For example, cerium, which is the most abundant in the lanthanide series, is more abundant that cobalt, tin and zinc; and thulium, which is the least abundant, occurs in similar quantities to mercury and is more abundant than silver. Although lanthanides are found in many ores, bastnaesite, monazite and xenotime are the major sources for industrial production. These ore deposits are located mainly in China, the USA, India, Australia and Brazil; China has the most important reserves. Separation of each element has been established on an industrial scale, and all the lanthanides except promethium are available as metal ingots or as powders, oxides and salts at moderate prices.

2.4 TOXICITY

According to some investigations on the toxicity of lanthanide salts, chlorides, nitrates and citrates of the lanthanide series have almost the same level of toxicity as sodium chloride (LD_{50} of 2000–7000 mg kg^{-1} in mice versus 4000 mg kg^{-1} for NaCl). These data indicate that lanthanides are orally non-toxic. However, moderate toxicity appears when they are introduced via the intraperitoneal route. Therefore, care must be taken not to introduce these salts into the lungs or eyes.

–3–

The Use of Lanthanide Metals in Synthesis

3.1 GENERAL ASPECTS

All lanthanide metals except radioactive promethium are produced on an industrial scale and they are commercially available in pure form as ingots or as powder. Their costs depend on their natural abundance. Lanthanum, cerium and neodymium can be purchased at moderate prices, and the price of samarium is not prohibitive. Other lanthanide metals are more expensive.

Lanthanide metals are utilized for the reduction of organic functional group or carbon–carbon bond-forming reactions. Metallic cerium, which is the cheapest of all the lanthanides, is potentially useful as a reducing agent. Samarium, europium and ytterbium, which are capable of forming a relatively stable divalent state, promote unique functional group transformations. Of these three metals, samarium is currently the most synthetically useful.

3.2 CERIUM

Metallic cerium is capable of reducing organic functional groups. For example, α-heterosubstituted ketones are readily reduced by cerium metal in the presence of acetic acid at room temperature [1]. Under similar conditions, the nitro group is reduced to an amino group in virtually quantitative yield. The reducing ability of cerium is greatly enhanced by pretreatment with mercury(II) chloride or iodine. Organic halides including fluoroalkanes are subjected to reductive coupling or reduction to saturated alkanes by this activated cerium. By the same reagent, ketones and imines are reductively coupled to 1,2-diols and 1,2-diamine derivatives, respectively [2].

The Barbier-type reaction and the Reformatsky-type reaction proceed in the presence of cerium. The chemoselectivity is noteworthy in these reactions. Thus, carbonyl groups are selectively subjected to the addition reaction, while ester, cyano and nitro groups remain unchanged under these conditions [3,4].

The reaction of ethyl 3-bromo- or 3-iodopropionic acid esters with cerium metal in THF produces cerium homoenolates [5]. The generated homoenolates react with carbonyl compounds to give γ-lactones as the major products.

$$X(CH_2)_2CO_2Et \xrightarrow[THF]{Ce} X_2Ce_{\text{mun}}\overset{O}{\underset{OEt}{\diagdown}} \xrightarrow{R^1COR^2} O \overset{R^1}{\underset{R^2}{\diagup}}$$

$$X = Br, I$$

Preparation of γ-lactones from ketones and ethyl 3-bromopropionate or ethyl 3-iodopropionate. General procedure [5]

Cerium powder (0.56 g, 4 mmol) and a trace of iodine (0.010 g, 0.04 mmol) were placed in a 50 ml two-necked round-bottomed flask equipped with a magnetic stirring bar. The flask was flushed with nitrogen several times. Ethyl 3-bromopropionate or ethyl 3-iodopropionate (4 mmol) in THF (2 ml) is added to cerium through a rubber septum with a syringe. The mild exothermic reaction started in a few minutes, and then ketone (4 mmol) in 3 ml of THF was injected into the solution. The mild exothermic reaction continued, and additional THF (5 ml) was introduced. The resulting solution was stirred at room temperature for 2 h, during which time the cerium powder was consumed almost completely and a fine black precipitate was formed. The mixture was treated with dilute HCl and extracted with ether (3 x 30 ml). The extract was washed with brine and dried over $MgSO_4$. The solvent was removed by evaporation, and the γ-lactone isolated by column chromatography on silica gel using hexane–ether (1 : 1) as the eluent.

REFERENCES

1. T. Imamoto, *Rev. Heteroatom Chem.* **3,** 87 (1990).
2. T. Imamoto, T. Kusumoto, Y. Hatanaka and M. Yokoyama, *Tetrahedron Lett.* **23,** 1353 (1982).
3. T. Imamoto, T. Kusumoto, Y. Tawarayama, Y. Sugiura, T. Mita, Y. Hatanaka and M. Yokoyama, *J. Org. Chem.* **49,** 3904 (1984).
4. S. Fukuzawa, T. Fujinami and S. Sakai, *J. Organometal. Chem.* **299,** 179 (1986).
5. S. Fukuzawa, N. Sumimoto, T. Fujinami and S. Sakai, *J. Org. Chem,* **55,** 1628 (1990).

3.3 SAMARIUM

3.3.1 Cyclopropanation of allylic alcohols

Samarium metal reacts with diiodomethane or chloroiodomethane in THF to generate divalent samarium species denoted as ICH_2SmI and $ClCH_2SmI$. These species decompose via α-elimination to give samarium(II) halide and ethylene. When the reactions are carried out in the presence of allylic alcohols, the reactive intermediates are trapped to provide cyclopropane derivatives [1,2]. The reaction is best carried out in the presence of catalytic amounts of mercury(II) chloride at −78 °C through to room temperature. Other 1,1-halides such as 1,1-diiodoethane can be employed to give the corresponding cyclopropane derivatives.

This cyclopropanation reaction is highly chemoselective. Thus, in contrast to the conventional cyclopropanation using zinc-copper couple, diethyl zinc or triisobutylaluminium, the olefinic moiety of allylic alcohols is selectively subjected to cyclopropanation leaving other olefinic groups unchanged.

In addition to the enhanced chemoselectivity, high diastereoselectivity is observed in this samarium-promoted cyclopropanation reaction. Representative examples are shown in Table 3.1. In most cases the reactions proceed via a hydroxy-directed mechanism. In the case of cyclic allyl alcohols, the methylenation occurs from the same side as the hydroxy group [1–4]. A high degree of stereoselectivity is observed also in the reactions of acyclic allylic alcohols [1,2,4–7]. The stereochemical outcome has been reasonably explained by Molander with the consideration of the Houk model and the coordination of the hydroxy group to the divalent samarium atom [1, 2].

Cyclopropanation of allylic alcohols using samarium and dihalomethane. General procedure [2]

To a dry, 25 ml round-bottomed flask equipped with a stirrer bar was added samarium metal (0.316 g, 2.1 mmol). The flask was flushed with argon and flamed dry. To the cooled flask was added 5 ml of THF, followed by a solution of mercuric chloride (0.054 g, 0.2 mmol) in 5 ml of THF. This was stirred for 10 min followed by addition of the allylic alcohol (0.5 mmol). The flask was cooled to −78 °C, and the dihalomethane (diiodomethane or chloroiodomethane) (2.0 mmol) was added dropwise. The mixture was allowed to warm to room temperature and was stirred for an additional 1–2 h. The reaction was followed by TLC or GC. The reaction was quenched with saturated K_2CO_3 and the mixture extracted with ether. The ether layer

TABLE 3.1
Samarium-promoted cyclopropanation of allylic alcohols

Allylic alcohol	1,1-Dihaloalkan	Product(s) (%)	Ref.
(SiMe₃ alkynyl cyclopentenol structure)	CH₂ClI	SiMe₃ —OH (77)	[3]
(cyclohexenol structure) OH	CH₂I₂	OH, H, H (92)	[1]
(cyclohexenol structure) OH	CH₃CHI₂	OH, H, CH₃, H, H (80); OH, H, H, CH₃, H (5)	[1]
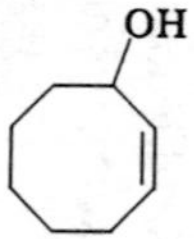 (cyclooctenol structure) OH	CH₂I₂	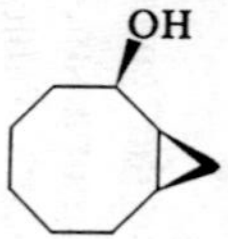 OH (>95% de) (85)	[2]

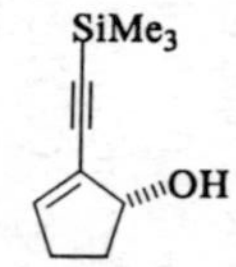

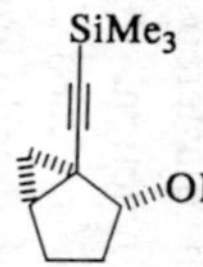

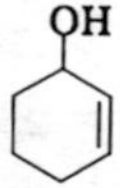

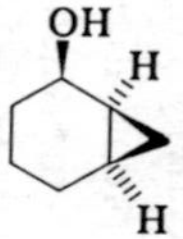

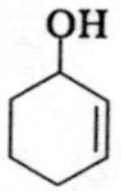

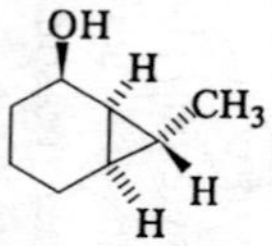

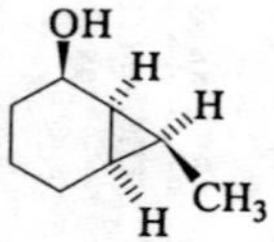

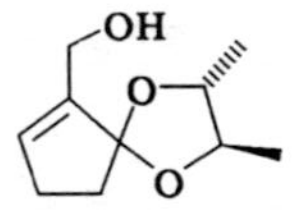

[4]

[2]

[2]

[5]

(94)

(99.5% de) (88)

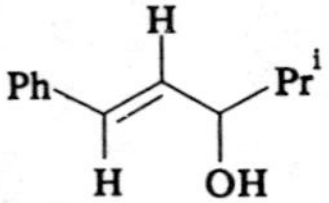

(99.5% de) (99)

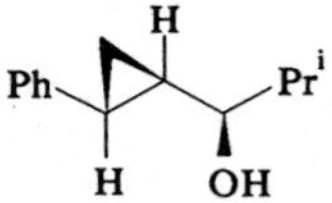

(92)

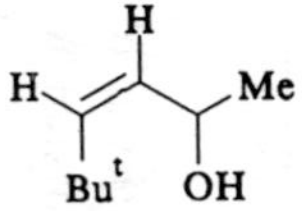

CH$_2$ClI

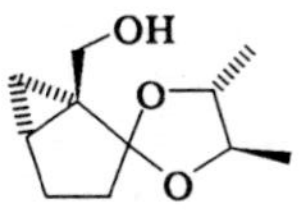

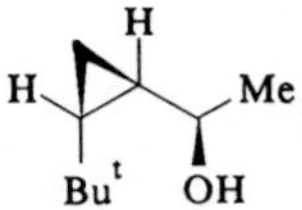

CH$_2$I$_2$

CH$_2$I$_2$

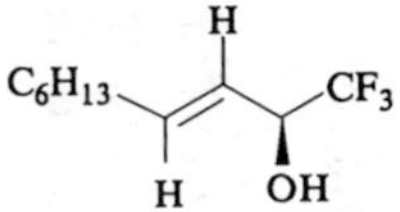

CH$_2$I$_2$

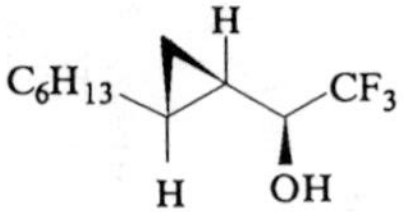

was washed with brine three times, dried over K_2CO_3, filtered, and concentrated *in vacuo* to yield the crude material. Flash chromatography or Kugelrohr distillation yielded the pure cyclopropyl carbinol.

3.3.2 Synthesis of cyclopropanols from ketones or esters

Metal enolates react with diiodomethane in the presence of samarium(II) iodide to afford cyclopropanol derivatives [8]. This cyclopropanation reaction permits synthesis of cyclopropanols from ketones via *in situ* generation of metal enolates and subsequent treatment with Sm–CH_2I_2 reagent systems. It is noted that α-haloketones and 1,4-diketones are converted to cyclopropanols in good yields [8]. These reaction sequences involve generation of samarium enolates and subsequent cyclopropanation reactions.

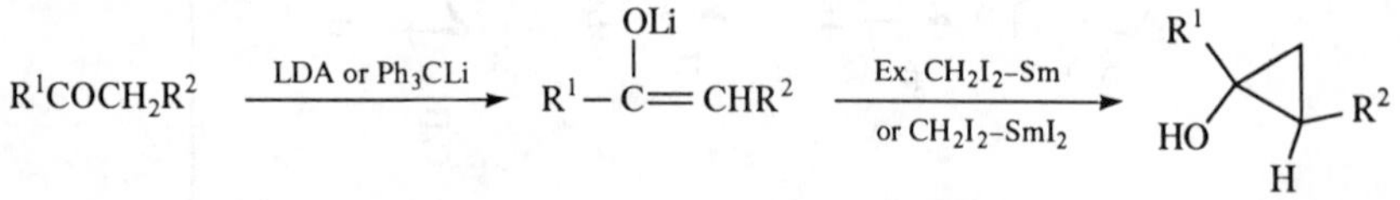

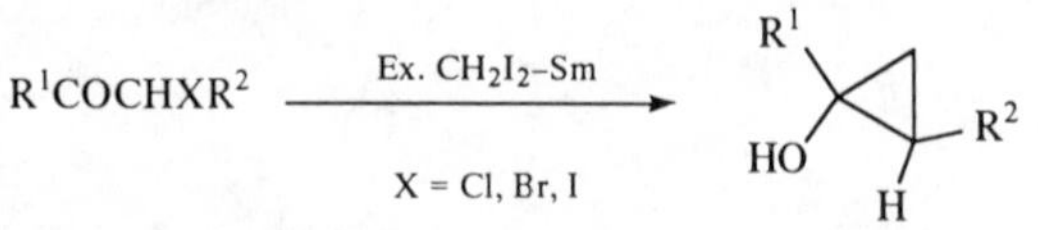

Direct conversion of esters to cyclopropanols via tandem one-carbon homologation can be achieved by the use of the CH_2I_2–Sm reagent system [9]. It is noted that the other olefinic moiety remains unchanged during the reaction.

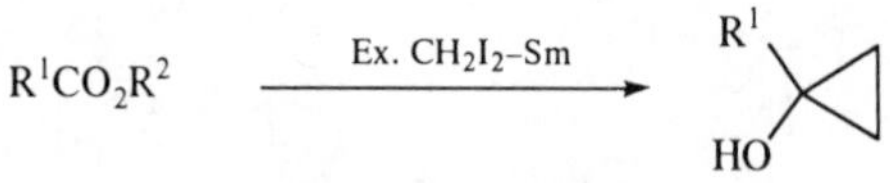

Preparation of (Z)-(1-hydroxycyclopropyl)heptadec-8-ene from methyl oleate [9]

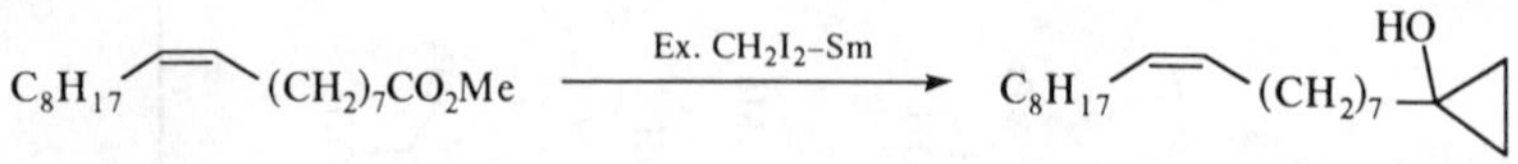

Samarium powder (0.60 g, 4 mmol) was placed in a 30 ml two-necked flask and covered with dry THF (2 ml). The mixture was warmed to 50 °C, and to this mixture was added a solution of methyl oleate (0.297 g, 1 mmol) and diiodomethane

(0.804 g, 3 mmol) in THF (8 ml) under argon. (The reaction usually begins a few minutes after addition of several drops of the solution.) The remainder of the solution was slowly added over 1.5 h with vigorous stirring. After this addition, the mixture was cooled, treated with 1 M HCl, and extracted with ether. The solvent was evaporated, and the residue chromatographed on silica gel using hexane–ethyl acetate (5 : 1) as the eluent to give (Z)-(1-hydroxycyclopropyl)heptadec-8-ene (0.206 g, 70%) as a colourless oil.

3.3.3 Iodomethylation of carbonyl compounds

Nucleophilic iodomethylation of carbonyl compounds is known to be difficult to achieve under ordinary conditions, because iodomethyl organometallic compounds are in general only stable at low temperature and generally decompose at elevated temperature via α-elimination. This kind of reaction proceeds smoothly at around room temperature using samarium metal or samarium(II) iodide [9–11].

$$
R^1COR^2 \xrightarrow[\text{THF, 0 °C to rt}]{CH_2I_2\text{-Sm or } CH_2I_2\text{-SmI}_2} R^1 \overset{\displaystyle OH}{\underset{\displaystyle R^2}{\overset{|}{\underset{|}{C}}}} \!-\! CH_2I
$$

Simple aliphatic ketones are converted into the corresponding iodohydrins in good to high yields. Easily enolizable ketones such as dibenzyl ketone and β-tetralone are also subjected to iodomethylation. However, the reactions of α,β-unsaturated carbonyl compounds and aromatic aldehydes result in low yields of the iodohydrins.

Reaction of cyclohexanone with diiodomethane in the presence of samarium; preparation of 1-iodomethylcyclohexanol [9]

Samarium powder (0.300 g, 2 mmol) was placed in a 20 ml two-necked flask with a dropping funnel containing a mixture of cyclohexanone (0.098 g, 1 mmol), diiodomethane (0.80 g, 3 mmol) and dry THF (6 ml). The THF solution was slowly added at 0 °C under argon. The reaction was typically initiated within a few minutes, with development of a characteristic deep-green colour, and the addition of the solution required about 20 min. After addition, stirring was continued for an additional 20 min at the same temperature. The reaction mixture was treated with 1 M HCl and extracted with ether. The combined extracts were washed with aqueous $Na_2S_2O_3$ and brine, dried ($MgSO_4$) and evaporated. The residue was chromatographed on silica gel using hexane–ethyl acetate (5 : 1) to give a pale-yellow solid (0.20 g, 84%). Recrystallization from hexane gave pure liodomethylcyclohexanol, m.p. 69–70 °C.

REFERENCES

1. G. A. Molander and J. B. Etter, *J. Org. Chem.* **52,** 3942 (1987).
2. G. A. Molander and L. S. Harring, *J. Org. Chem.* **54,** 3525 (1989).
3. D. L. J. Clive and S. Daigneault, *J. Org. Chem.* **56,** 3801 (1991).
4. M. Kabat, J. Kiegiel, N. Cohen, K. Toth, P. M. Wovkulich and M. R. Uskokovic, *Tetrahedron Lett.* **32,** 2343 (1991).
5. T. Yamazaki, J. T. Lin, M. Takeda and T. Kitazume, *Tetrahedron Asymm.* **1,** 351 (1990).
6. K. Shimamoto, M. Ishida, H. Shinozaki and Y. Ohfune, *J. Org. Chem.* **56,** 4167 (1991).
7. M. Lautens and P. H. M. Delanghe, *J. Org. Chem.* **57,** 798 (1992).
8. T. Imamoto and N. Takiyama, *Tetrahedron Lett.* **28,** 1307 (1987).
9. T. Imamoto, T. Hatajima, N. Takiyama, T. Takeyama, Y. Kamiya and T. Yoshizawa, *J. Chem. Soc. Perkin Trans.* **1,** 3127 (1991).
10. T. Imamoto, T. Takeyama and H. Koto, *Tetrahedron Lett.* **27,** 3243 (1986).
11. T. Tabuchi, J. Inanaga and M. Yamaguchi, *Tetrahedron Lett.* **27,** 3891 (1986).

3.4 YTTERBIUM

3.4.1 Reduction of organic functional groups by ytterbium

Ytterbium metal dissolves in liquid ammonia forming a blue solution containing the ammoniated electron. This solution containing alcohol as proton source reduces aromatic systems to 1,4-dihydroaromatics, α,β-unsaturated carbonyl compounds to saturated ketones, and alkynes to *trans* alkenes [1].

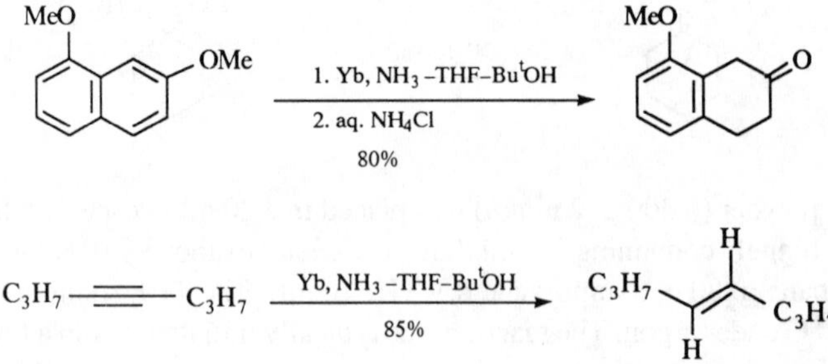

This reagent system represents a useful reducing agent which resembles the Birch reaction using alkali metal. The fact that strongly basic hydroxide is avoided during work-up lends this method certain advantages over the more reactive metals commonly used in electron-transfer chemistry.

Reduction of organic functional groups by ytterbium in organic medium has also been studied. Olefinic double bonds and imino groups are also reduced by ytterbium in methanol to give alkanes and secondary amines, respectively; however, the substrates are limited to compounds in which the double bond is conjugated to an aromatic group [2,3].

Reduction of stilbene by ytterbium in methanol-d₄; preparation of 1,2-dideuterio-1,2-diphenylethane [3]

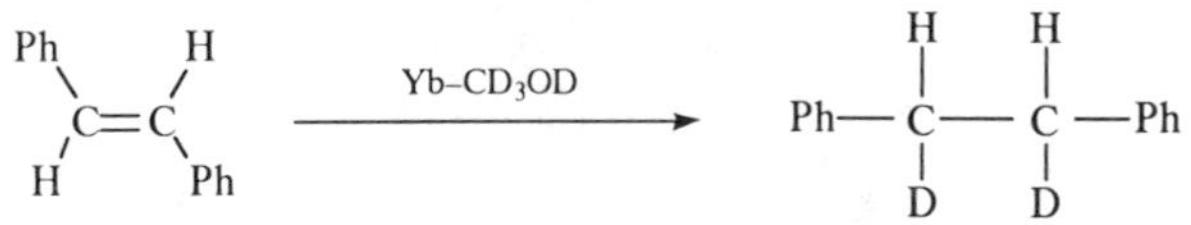

A mixture of (*E*)-stilbene (0.090 g, 0.5 mmol), CD₃OD (0.5 ml) and THF (2 ml) was added to a mixture of activated ytterbium (*ca.* 40 mesh) (0.26 g, 1.5 mmol) and dry THF (2 ml) under nitrogen. After being stirred for 18 h, the resulting dark slurry was treated with 2 M HCl and extracted with ether. The combined extracts were washed with brine and dried over Na_2SO_4. The solvent was evaporated, and the residue chromatographed on silica gel to give 1,2-dideuterio-1,2-diphenylethane (0.087 g, 95%).

3.4.2   Carbon–carbon bond-forming reactions promoted by ytterbium

The reaction of diaryl ketones with ytterbium occurs readily at room temperature to give oxymetallocyclic compounds as reactive intermediates [4,5]. The carbonyl groups in diaryl ketones are "umpoled" by this reaction from being electrophilic to being nucleophilic in character. Thus, the ytterbium organometallics react with various electrophiles to give a variety of coupling products such as unsymmetrical 1,2-diols, 1,3-diols and α-hydroxycarbonyl compounds [4–6].

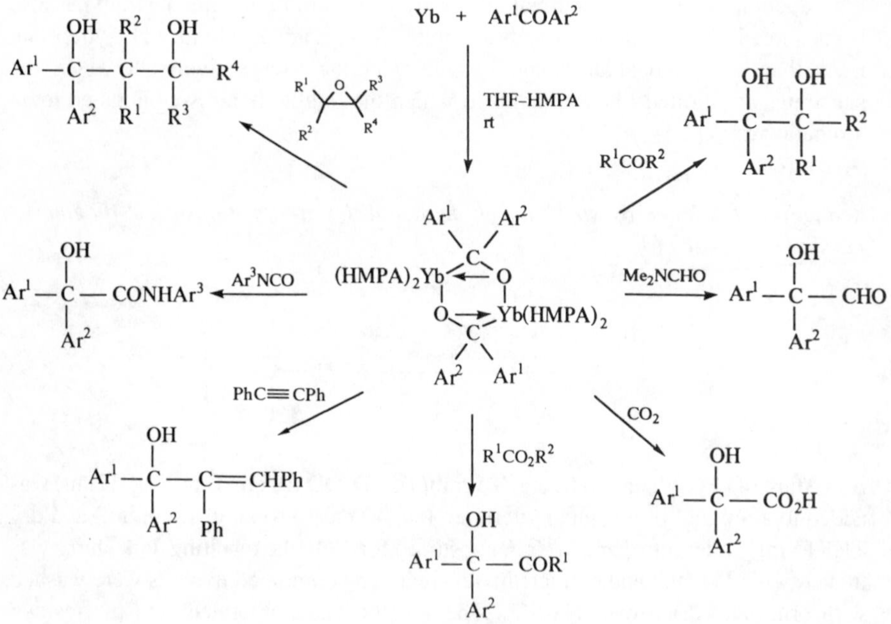

$$Yb \;+\; Ar^1COAr^2$$

$$\xrightarrow[\text{rt}]{\text{THF–HMPA}}$$

Central intermediate:

$$\begin{array}{c} Ar^1 \quad Ar^2 \\ \diagdown \;\diagup \\ C \\ (HMPA)_2Yb \leftarrow O \\ O \rightarrow Yb(HMPA)_2 \\ C \\ \diagup\;\diagdown \\ Ar^2 \quad Ar^1 \end{array}$$

Top-left product (from epoxide R^1,R^2,R^3,R^4 / O):

$$\begin{array}{c} \text{OH}\quad R^2\quad \text{OH} \\ | \qquad | \qquad | \\ Ar^1 - C - C - C - R^4 \\ | \qquad | \qquad | \\ Ar^2 \quad R^1 \quad R^3 \end{array}$$

Top-right product (with R^1COR^2):

$$\begin{array}{c} \text{OH}\quad \text{OH} \\ | \qquad | \\ Ar^1 - C - C - R^2 \\ | \qquad | \\ Ar^2 \quad R^1 \end{array}$$

Left (with Ar^3NCO):

$$\begin{array}{c} \text{OH} \\ | \\ Ar^1 - C - CONHAr^3 \\ | \\ Ar^2 \end{array}$$

Right (with Me_2NCHO):

$$\begin{array}{c} \text{OH} \\ | \\ Ar^1 - C - CHO \\ | \\ Ar^2 \end{array}$$

Lower-left (with $PhC\equiv CPh$):

$$\begin{array}{c} \text{OH} \\ | \\ Ar^1 - C - C = CHPh \\ | \qquad | \\ Ar^2 \quad Ph \end{array}$$

Bottom (with $R^1CO_2R^2$):

$$\begin{array}{c} \text{OH} \\ | \\ Ar^1 - C - COR^1 \\ | \\ Ar^2 \end{array}$$

Lower-right (with CO_2):

$$\begin{array}{c} \text{OH} \\ | \\ Ar^1 - C - CO_2H \\ | \\ Ar^2 \end{array}$$

The reaction of aromatic imines with ytterbium has also been studied. Umpolung of imine functionality is observed, and the subsequent reactions with carbon dioxide provide amino acid derivatives [7].

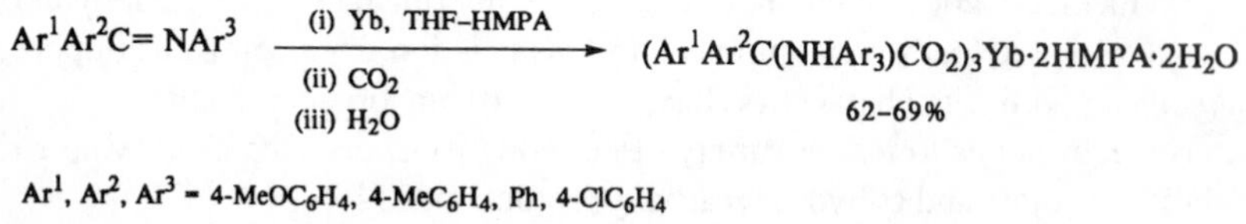

$$Ar^1Ar^2C=NAr^3 \quad \xrightarrow[\substack{\text{(ii) } CO_2 \\ \text{(iii) } H_2O}]{\text{(i) Yb, THF–HMPA}} \quad (Ar^1Ar^2C(NHAr_3)CO_2)_3Yb\cdot 2HMPA\cdot 2H_2O$$

$$62\text{–}69\%$$

Ar^1, Ar^2, Ar^3 = 4-MeOC$_6$H$_4$, 4-MeC$_6$H$_4$, Ph, 4-ClC$_6$H$_4$

On the other hand, the reaction of α,β-unsaturated carbonyl compounds with ytterbium metal gives cyclodimerization products in good yields [8–10].

*Preparation of 2-benzoyl-1,3,4-triphenylcyclopentanols (**1** and **2**) [11]*

(1)   (2)
68%   27%

A 200 ml two-necked, round-bottomed flask equipped with magnetic stirring bar, septum cap and gas inlet was charged with 1.73 g (10 mmol) of ytterbium metal (*ca.* 40 mesh). The apparatus was evacuated and filled with argon and a positive pressure was maintained with an argon-filled balloon on a T-tube. Iodomethane (*ca.* 5 μl) was added and the flask was warmed by a dryer for about 30 min to activate the ytterbium metal. A solution of chalcone (2.08 g, 10 mmol) in THF (40 ml) and HMPA (CAUTION – CANCER SUSPECT AGENT) (10 ml) was added through the septum cap, and the mixture was stirred at room temperature for 4 h. The reaction mixture was treated with 2 M HCl (20 ml) and extracted with ether. The combined extracts were washed with brine and dried over Na_2SO_4. The solvent was evaporated and the residue chromatographed on silica gel using hexane–ethyl acetate (6 : 1) as the eluent to give compounds **1** (1.42 g, 68%), m.p. 176–179 °C, and **2** (0.56 g, 27%), m.p. 206–212 °C.

REFERENCES

1. J. D. White and G. L. Larson, *J. Org. Chem.* **43**, 4555 (1978).
2. Z. Hou, H. Taniguchi and Y. Fujiwara, *Chem. Lett.* 305 (1987).
3. K. Takai, Y. Tsubaki, S. Tanaka, F. Beppu and Y. Fujiwara, *Chem. Lett.* 203 (1990).
4. Z. Hou, K. Takamine, O. Aoki, H. Shiraishi, Y. Fujiwara and H. Taniguchi, *J. Org. Chem.* **53**, 6077 (1988).
5. Z. Hou, H. Yamazaki, Y. Fujiwara and H. Taniguchi, *Organometallics* **11**, 2711 (1992).
6. K. Takaki, S. Tanaka, F. Beppu, Y. Tsubaki and Y. Fujiwara, *Chem. Lett.* 1427 (1990).
7. K. Takai, S. Tanaka and Y. Fujiwara, *Chem. Lett.* 493 (1991).
8. K. Takaki, F. Beppu, S. Tanaka, Y. Tsubaki, T. Jintoku and Y. Fujiwara, *J. Chem. Soc. Chem. Commun.* 516 (1990).
9. K. Takaki, K. Nagase, F. Beppu and Y. Fujiwara, *Chem. Lett.* 1665 (1991).
10. K. Takaki, K. Nagase, F. Beppu, T. Shindo and Y. Fujiwara, *Chem. Lett.* 1669 (1991).
11. Y. Fujiwara and K. Takaki, in *Practical Organometallic Chemistry for Synthetic Chemists*, ed. F. Sato, K. Yamamoto and T. Imamoto, p. 251. Kodansha, Tokyo, 1992 (reproduced by permission from Y. Fujiwara and Kodansha).

3.5 ALLOYS AND ACTIVATED METALS

3.5.1 Alloys

Lanthanide intermetallics (alloys) such as $LaNi_5$, $PrCo_5$ and $SmCo_5$ absorb large quantities of hydrogen rapidly and reversibly under mild conditions. For example, $LaNi_5$ absorbs hydrogen to form $LaNi_5H_6$. These unique properties of the alloys can be utilized for the hydrogenation of unsaturated organic compounds. Alkynes, alkenes, aldehydes, ketones, nitriles, imines and nitro compounds are hydrogenated in excellent yields with $LaNi_5H_6$ [1].

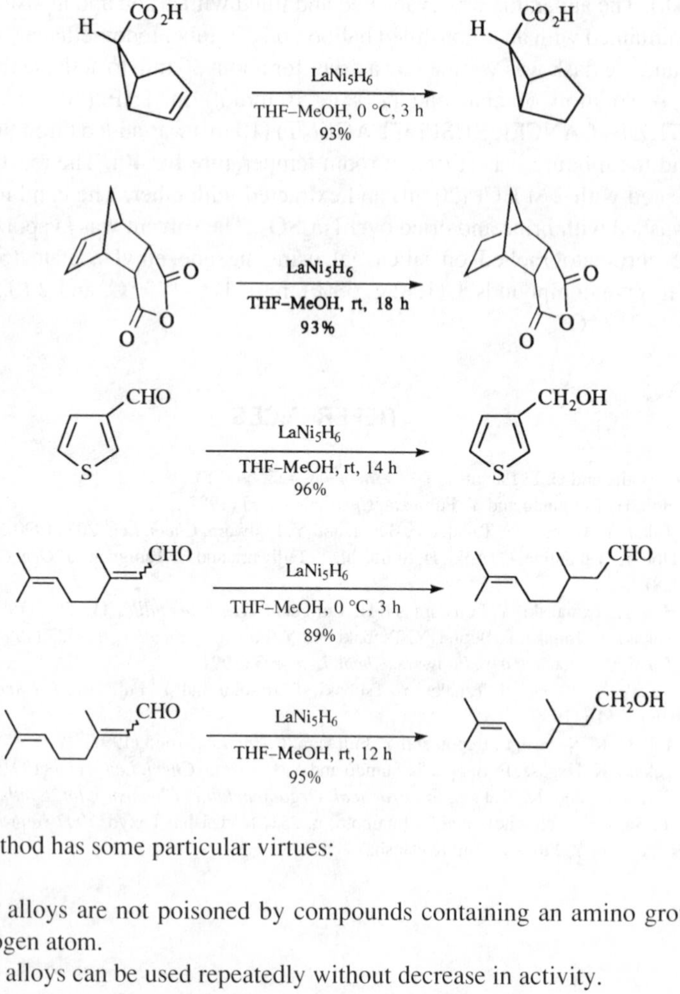

This method has some particular virtues:

(1) The alloys are not poisoned by compounds containing an amino group or a halogen atom.
(2) The alloys can be used repeatedly without decrease in activity.

(3) The reaction conditions are mild, and selective hydrogenations of organic functional groups can be achieved.

Several kinds of lanthanide intermetallics are commercially available from Aldrich and can be employed for these reductions.

Hydrogenation with LaNi$_5$H$_6$. General procedure [1]

A 20 ml long-necked flask containing LaNi$_5$ (3 g) was placed in an autoclave. The system was evacuated (*ca.* 0.1 mmHg) and the autoclave heated to 200 °C. After 10 min at this temperature, the autoclave was cooled to room temperature. This operation was repeated five times. (If one uses the recovered alloy, the repetition of this operation is not required.) Then, the autoclave was immersed in an ice bath and finally in a dry ice–acetone bath. Excess hydrogen was released, and the flask was taken out in a stream of nitrogen and set on a magnetic stirrer.

A solution of organic compound (1 mmol) in THF–MeOH (2 : 3) (5 ml) was added using a syringe at –78 °C. The dry ice bath was then removed, and the mixture stirred under nitrogen at 0 °C through to room temperature. After completion of the reduction (checked by TLC), the mixture was filtered by suction to recover the alloy. The filtrate was concentrated, and the residue purified by chromatography on silica gel.

3.5.2 Activated metals

Vaporized samarium is highly reactive. It reacts readily with penta-methylcyclopentadiene to give bis(pentamethylcyclopentadienyl) samarium(II). The activated samarium catalyses hydrogenation of alkenes and alkynes [2,3].

REFERENCES

1.   T. Imamoto, T. Mita and M. Yokoyama, *J. Org. Chem.* **52,** 5695 (1987).
2.   W. J. Evans, I. Bloom and S. C. Engerer, *J. Catalysis* **84,** 468 (1983).
3.   H. Imamura, K. Kitajima and S. Tsuchiya, *J. Chem. Soc. Faraday Trans. 1* **85,** 1647 (1989).

$-4-$

Divalent Lanthanides

4.1 GENERAL ASPECTS

Europium, samarium and ytterbium form a relatively stable dipositive state, having half-filled ($Eu^{2+}=4f^7$), nearly half-filled ($Sm^{2+}=4f^6$) and filled ($Yb^{2+}=4f^{14}$) 4f orbitals, respectively. Oxidation-reduction potentials of Ln^{3+}/Ln^{2+} in aqueous medium are -0.35 V, -1.55 V and -1.15 V for europium, samarium and ytterbium. Therefore, the reducing power is predicted to be the following order: Eu(II) < Yb(II) < Sm(II). Europium(II) is the most stable and exists even in aqueous medium. Divalent samarium is highly reactive and is capable of reducing organic functional groups more readily than ytterbium and europium. Since samarium is also the cheapest of these three lanthanides, samarium(II) has become an important reducing agent.

Samarium(II) iodide (SmI_2), dicyclopentadienylsamarium(II) (($C_5H_5)_2Sm$) and bis(pentamethylcyclopentadienyl)samarium(II) (($C_5Me_5)_2Sm$) are representative reagents of divalent samarium. In particular, SmI_2 is the most useful and frequently employed in organic synthesis.

4.2 PREPARATION OF LANTHANIDE(II) REAGENTS

4.2.1 Samarium(II) reagents

4.2.1.1 Samarium(II) iodide

Samarium(II) iodide is readily prepared by the reaction of samarium metal with 1,2-diiodoethane [1,2], diiodomethane [2], iodine [3] or mercury(II) iodide [4] in THF. It is usually prepared by the use of 1,2-diiodoethane or diiodomethane; diiodomethane is cheaper than the former and is recommended.

Evaporation of the solvent gives a brown powder whose crystal structure has not yet been determined. Samarium(II) iodide is usually used as a THF solution for organic synthesis. The solution can be stored under argon or nitrogen for a long time at room temperature, and it is commercially available from Aldrich.

Preparation of a solution of SmI$_2$ in THF [1]

Samarium powder (*ca.* 40 mesh, 3 g, 0.02 mol) was placed in a 200 ml two-necked flask fitted with a dropping funnel containing 1,2-diiodoethane (2.82 g, 0.01 mol) or diiodomethane (2.68 g, 0.01 mol) in 100 ml of dry THF. About 5 ml of the THF solution was added and the content of the flask was magnetically stirred. After a few minutes, the reaction was initiated with appearance of a blue-green colour. The flask was immersed in an ice bath, and the remainder of the THF solution slowly added with stirring over 1 h. After this addition, the mixture was stirred vigorously for 5 h at room temperature to give a 0.1 M solution of SmI$_2$.

4.2.1.2 Dicyclopentadienylsamarium(II) and bis(pentamethylcyclopentadienyl)samarium(II)

Dicyclopentadienylsamarium(II), $(C_5H_5)_2Sm$, is prepared as a red powder by the reaction of SmI$_2$ with sodium cyclopentadienide in THF under argon at room temperature [5]. This reagent is insoluble in most organic solvents, but promotes several types of organic transformations.

Bis(pentamethylcyclopentadienyl)samarium(II), $(C_5Me_5)_2Sm$, is prepared by the reaction of SmI$_2$ with two equivalents of KC$_5$Me$_5$ [6–8] or reaction of vaporized samarium with 1,2,3,4,5-pentamethyl-1,3-cyclopentadiene (C$_5$HMe$_5$) [9]. This compound is soluble in THF and exhibits intriguing reactivities different from SmI$_2$ and $(C_5H_5)_2Sm$.

Preparation of dicyclopentadienylsamarium(II) [5]

Samarium diiodide (0.1 M in THF, 60 ml, 6 mmol) was slowly added to 0.4 M solution of sodium cyclopentadienide in THF (30 ml, 12 mmol). A dark-purple precipitate formed immediately, and was decanted within 1 h. The precipitate was washed twice with THF to remove sodium iodide.

Dicyclopentadienylsamarium(II) can be stored for a few days in a Schlenk tube in THF without decomposition.

Preparation of bis(pentamethylcyclopentadienyl)samarium(II) [6,7]

In a glove box, KC$_5$Me$_5$ (5.43 g, 31.2 mmol) was added to a stirring solution of SmI$_2$ (THF)$_2$ (7.78 g, 14.2 mmol) in 75 ml of THF in a 125 ml Erlenmeyer flask. The colour of the solution rapidly changed from blue-green to purple as an off-white solid (KI) was formed. After 4 h at ambient temperature, the THF was removed by rotary evaporation, and 100 ml of toluene was added. The resulting mixture was stirred vigorously for 10 h and then filtered. The solvent was removed from the filtrate by rotary evaporation, leaving solid $(C_5Me_5)_2Sm(THF)_n$, where $1 \leqslant n \leqslant 2$. The degree of solvation was conveniently monitored by integration of the absorptions in the NMR spectrum in benzene-d$_6$. By dissolving this solid in THF followed by removal of the solvent by rotary evaporation, the disolvate $(C_5Me_5)_2Sm(THF)_2$ was formed (5.95 g, 74%). Recrystallization from THF (solution saturated at 30 °C cooled to –25 °C overnight) gave large, purple crystals (5.52 g in two crops, 69%). The monosolvate, $(C_5Me_5)_2Sm(THF)$, could be obtained by repeated evaporation of the original toluene extraction solutions. THF-free complex was obtained by sublimation of $(C_5Me_5)_2Sm(THF)_2$ at 85–125 °C under vacuum (2 x 10^{-5} mmHg).

4.2.2 Europium(II) and ytterbium(II) reagents

Europium(II) iodide [8] and ytterbium(II) iodide [1] are obtained by the reaction of europium and ytterbium with 1,2-diiodoethane in THF in a procedure similar to the preparation of SmI_2. In the same way, ytterbium(II) bromide can be prepared by the use of 1,2-dibromoethane [8]. Bis(pentamethylcyclopentadienyl)europium [7,8,10] and bis(pentamethylcyclopentadienyl)ytterbium(II) [8] are obtained in a procedure similar to the preparation of $(C_5Me_5)_2Sm$.

REFERENCES

1. P. Girard, J. L. Namy and H. B. Kagan, *J. Am. Chem. Soc.* **102**, 2693 (1980).
2. J. L. Namy, P. Girad and H. B. Kagan, *Nouv. J. Chem.* **5**, 479 (1981).
3. T. Imamoto and M. Ono, *Chem. Lett.* 501 (1987).
4. G. B. Deacon and C. M. Forsyth, *Chem. Lett.* 837 (1989).
5. J. L. Namy, J. Collin, J. Zhang and H. B. Kagan, *J. Organometal. Chem.* **328**, 81 (1987).
6. W. J. Evans, J. W. Grate, H. W. Choi, I. Bloom, W. E. Hunter and J. L. Atwood, *J. Am. Chem. Soc.* **107**, 941 (1985).
7. W. J. Evans, L. A. Hughes and T. P. Hanusa, *Organometallics* **5**, 1285 (1986).
8. P. L. Watson, T. H. Tulip and I. Williams, *Organometallics* **9**, 1999 (1990).
9. W. J. Evans, I. Bloom, W. E. Hunter and J. L. Atwood, *Organometallics* **4**, 112 (1985).
10. T. D. Tilley, R. A. Andersen, B. Spencer, H. Ruben, A. Zalkin and D. H. Templeton, *Inorg. Chem.* **19**, 2999 (1980).

4.3   REDUCTION OF ORGANIC FUNCTIONAL GROUPS BY SAMARIUM(II) IODIDE

Many organic functional groups are reduced by SmI_2 in the presence or absence of a proton source. The reduction rate largely depends on solvents or additives. Addition of HMPA as a cosolvent dramatically enhances the reducing power of SmI_2; various organic functional groups are reduced much more rapidly in HMPA–THF than in THF alone. This powerful reducing ability is ascribed to the strong electron donation of HMPA to the samarium ion, by virtue of complexation with enhancement of the electron releasing power of Sm^{2+}. Stabilization of the resulting Sm^{3+} ion by coordination is another driving force for the reduction. Dimethylaminourea and related compounds also elevate the reducing ability of Sm^{2+}, though they are not as effective as HMPA. The strong reducing ability of samarium(II) iodide is observed even in the presence of alkali metal hydroxide or phosphoric acid, although the conditions are rather drastic and are of limited applicability.

4.3.1 Organic halides and related compounds

Primary bromides and iodides are reduced by SmI_2 in dry THF at reflux to the corresponding alkanes in high yields [1]. No Wurtz coupling products were produced. Alkyl tosylates are also reduced under similar conditions. The reduction presumably proceeds via initial conversion to alkyl iodides, followed by reduction.

In contrast to the alkyl halides, reactions of benzylic or allylic halides with SmI_2 result in high yields of coupling products at room temperature [1]. The couplings preferentially occur in a head-to-head manner in most cases.

The reducing power of SmI_2 is dramatically increased by the addition of HMPA as the cosolvent [2,3]. Aliphatic, vinylic and aromatic iodides, bromides and chlorides are rapidly reduced to the corresponding hydrocarbons by SmI_2 in THF–HMPA (Table 4.1). Fluorides are not reduced under these conditions. Reduction of alkyl fluorides by SmI_2, however, proceeds with tungsten-lamp irradiation [4].

$$RX \quad \xrightarrow[\text{THF–HMPA}]{SmI_2} \quad RH$$

$$X = I, Br, Cl$$

From a deuterium incorporation study, it is suggested that aliphatic halides are reduced to organosamarium(III) intermediates, whereas aromatic halides accept one electron and the resulting radical species abstract hydrogen from the solvent THF. Reduction of optically active alkyl halides with SmI_2 results in racemization of products [5].

β-Chloro ethers undergo reductive 1,2-eliminations with SmI_2. The reactions of 3-chloro 2-substituted furans and 3-chloro 2-substituted pyrans provide either (E)- or (Z)-alkenes with excellent selectivity. This method is a new entry for the stereoselective synthesis of alkenes possessing a hydroxy group.

TABLE 4.1

Reduction of Organic Halides by SmI_2 in THF–HMPA [2]

Organic halides	Proton source	Conditions	Yield of product (%)
$C_{10}H_{21}I$	Pr^{i}OH	rt, 5 min	>95
$C_{10}H_{21}Br$	Pr^{i}OH	rt, 10 min	>95
$C_{10}H_{21}Cl$	Pr^{i}OH	60 °C, 8 h	>95
Iodocyclodecane	Pr^{i}OH	rt, 10 min	>95
Bromocyclodecane	Pr^{i}OH	rt, 10 min	>95
Cholesteryl chloride	Pr^{i}OH	rt, 3 h	99
1-Bromoadamanthane	Pr^{i}OH	rt, 10 min	>95
1-Iodonaphthalene	none	rt, 1 min	>95
1-Bromonaphthalene	none	rt, 5 min	98
1-Chloronaphthalene	none	rt, 15 min	>95
2-Bromobenzyl acetate	none	rt, 2 h	97

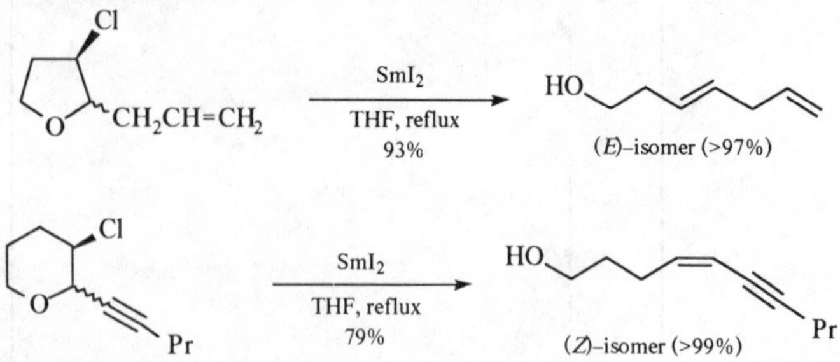

Similar reactions are utilized for the removal of protecting groups such as 2-chloroethyl carbamate and (2,2,2-trichloroethoxy)methoxy ether derivatives [6,7].

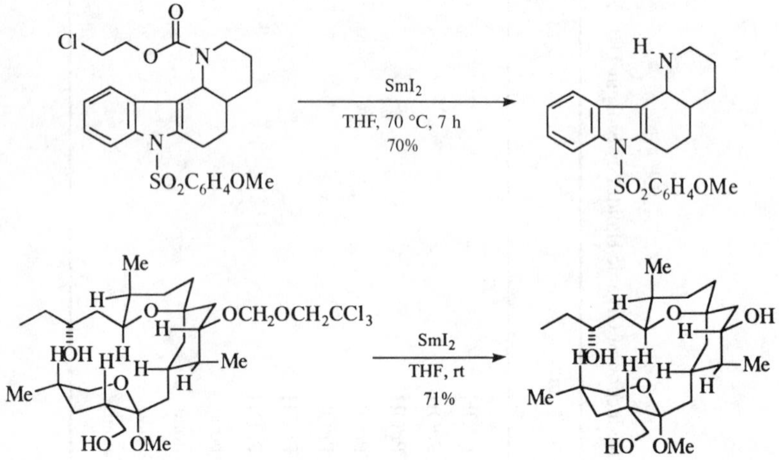

REFERENCES

1. P. Girard, J. L. Namy and H. B. Kagan, *J. Am. Chem. Soc.* **102**, 2693 (1980).
2. J. Inanaga, M. Ishikawa and M. Yamaguchi, *Chem. Lett.* 1485 (1987).
3. J. Inanaga, *Rev. Heteroatom Chem.* **3**, 75 (1990).
4. T. Imamoto and T. Takeyama, unpublished results.
5. H. M. Walborsky and M. Topolski, *J. Org. Chem.* **57**, 370 (1992).
6. T. P. Ananthanarayan, T. Gallagher and P. Magnus, *J. Chem. Soc. Chem. Commun.* 709 (1982).
7. D. A. Evans, S. W. Kaldor, T. K. Jones, J. Clardy and T. J. Staut, *J. Am. Chem. Soc.* **112**, 7001 (1990)

4.3.2 α-Heterosubstituted carbonyl and related compounds

α-Heterosubstituted ketones are reduced to unsubstituted ketones by treatment with SmI_2 in the presence of a proton source [1]. This SmI_2-promoted reduction has a wide range of applicability. Not only α-halo ketones, keto sulphides, sulphoxides and sulphones but also α-oxygenated ketones are reduced by SmI$_2$. The reductions are usually carried out at −78 °C under essentially neutral conditions, and hence other functional groups are as primary iodide, ester and ketone are tolerated under these conditions. Another characteristic feature of this method is that sterically hindered α-heterosubstituted ketones are reduced in synthetically useful yields. It is also noted that optically active β-hydroxy ketones can be synthesized from α,β-epoxy ketones without any loss of optical purity [2].

$$R^1-\underset{\underset{R^2}{|}}{\overset{\overset{X}{|}}{C}}-COR^3 \quad \xrightarrow[\text{THF-MeOH, −95 °C to −78 °C}]{\text{SmI}_2} \quad R^1-\underset{\underset{R^2}{|}}{\overset{\overset{H}{|}}{C}}-COR^3$$

X = Cl, Br, SR, S(O)R, SO$_2$R, HgCl, OR, OCOR, OH

α-Heterosubstituted esters are also reduced by SmI$_2$, though the reactions require more forcing conditions. Addition of HMPA, ethylene glycol, N,N-dimethylaminoethanol (DMAE) or pivalic acid as a cosolvent or an additive is quite effective; even hydroxy esters are reduced in high yields [3,4].

$$R^1-\underset{\underset{R^2}{|}}{\overset{\overset{X}{|}}{C}}-CO_2R^3 \quad \xrightarrow[\text{THF-(HMPA), −78 °C to rt}]{\text{SmI}_2} \quad R^1-\underset{\underset{R^2}{|}}{\overset{\overset{H}{|}}{C}}-CO_2R^3$$

X = Cl, Br, OR, OCOR, OH

Table 4.2 shows some representative examples of SmI$_2$-promoted reductions of α-heterosubstituted ketones and esters. This protocol is particularly useful for the deoxygenation of α-oxygenated carbonyl and related compounds, and it has widespread synthetic applicability.

Reduction of 2-acetoxy-5-iodo-1-phenyl-1-pentanone [1]

$$\text{PhCOCH(OAc)(CH}_2)_3\text{I} \quad \xrightarrow[\text{THF-MeOH, −78 °C}]{\text{SmI}_2} \quad \text{PhCO(CH}_2)_4\text{I}$$

To a slurry of samarium powder (0.32 g, 2.1 mmol) in THF (2 ml) at room temperature was added a solution of 1,2-diiodoethane (0.56 g, 2.0 mmol) in THF (2 ml). The resultant olive-green slurry was stirred at ambient temperature for 1 h, after which time the resulting dark-blue slurry of SmI$_2$ formed was cooled to −78 °C (dry ice/acetone) and treated with a solution of 2-acetoxy-5-iodo-1-phenyl-1-

TABLE 4.2

Reduction of α-Heterosubstituted Ketones and Esters by SmI$_2$

Substrate	Conditions	Product (%)	Ref.
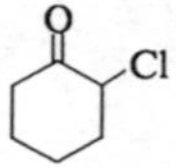	THF–MeOH, −78 °C	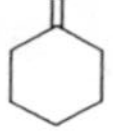 (100)	[1]
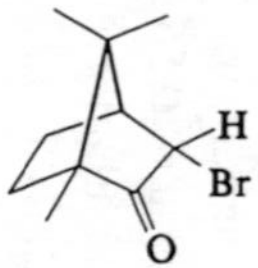	THF–MeOH, −78 °C	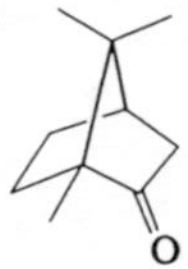 (85)	[1]
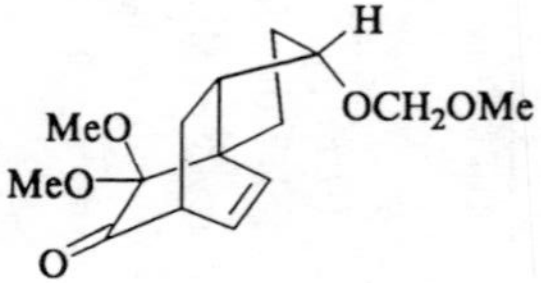	THF–MeOH, −78 °C	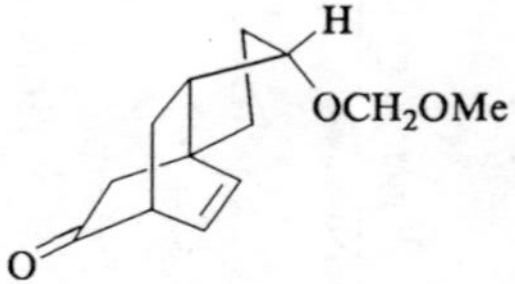 OCH$_2$OMe (92)	[5]
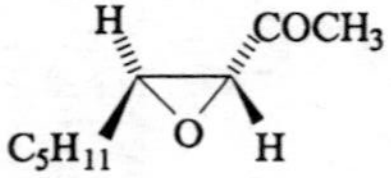	THF–MeOH, −90 °C	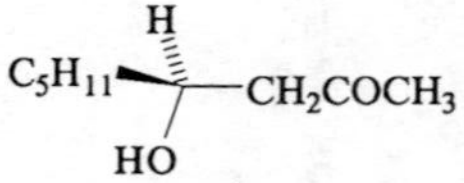 CH$_2$COCH$_3$ (94)	[2]
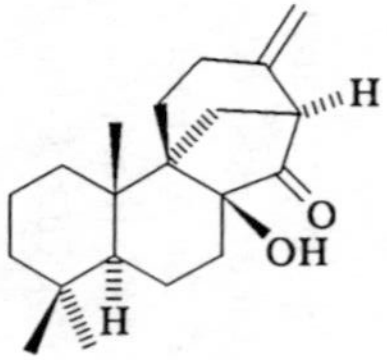	THF–Bu^{t}OH, 25 °C, 12 h	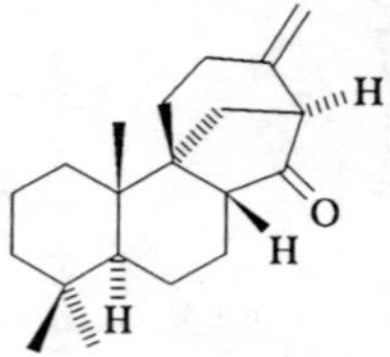 (87)	[6]

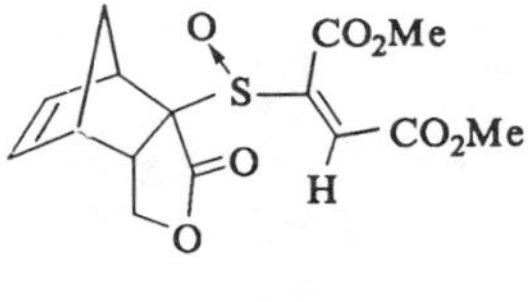

THF–Bu^{t}OH, rt, 10 min   (98)   [7]

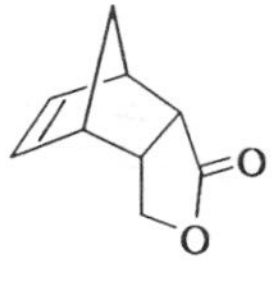

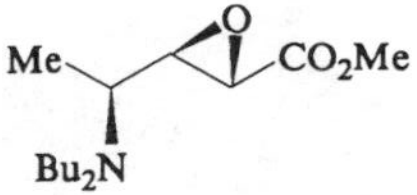

THF, –78 °C to rt, 2 h   (70)   [8]

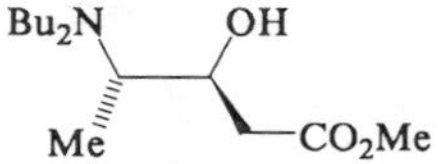

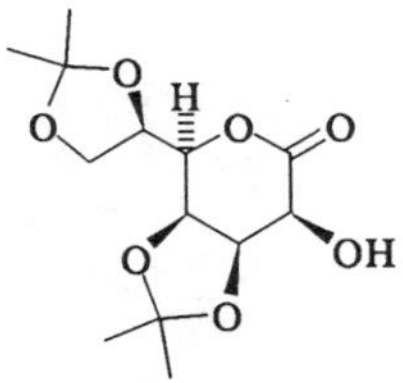

THF–HMPA–HO(CH$_2$)$_2$OH, rt, 30 min   (99)   [9]

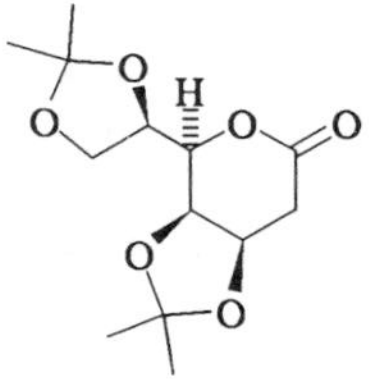

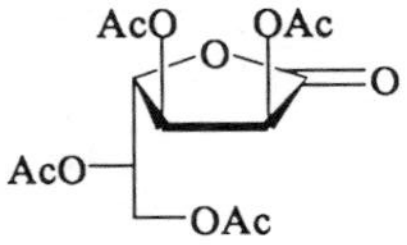

THF–Bu^{t}CO$_2$H, rt, 20 min   (67)   [10]

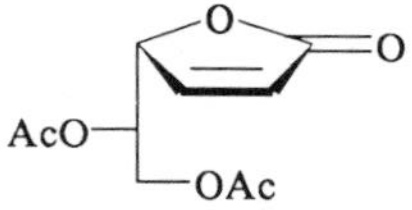

pentanone (0.35 g, 1 mmol) in MeOH (1 ml) and THF (2 ml). The resultant brown mixture was treated for 10 min at –78 °C, warmed to room temperature, and then poured into saturated aqueous K_2CO_3. The aqueous phase was extracted with ether (5 × 10 ml) and the combined extracts dried over $MgSO_4$. Solvent was evaporated and the residue was recrystallized from ether to afford 5-iodo-1-phenyl-1-pentanone (0.24 g, 87%), m.p. 72–73 °C.

Reduction of (R,R)-diisopropyl tartrate to (R)-diisopropyl malate [4]

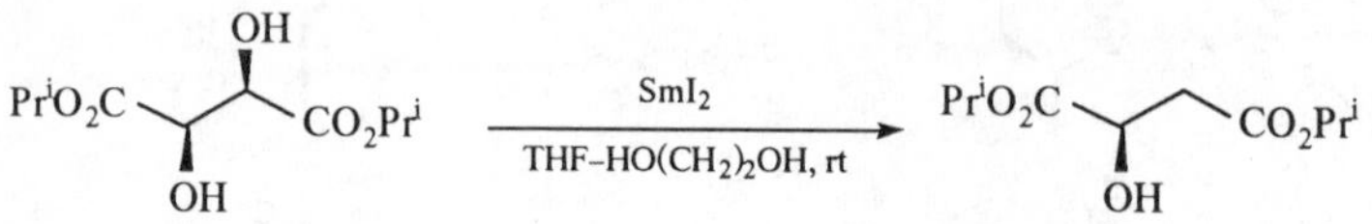

To a mixture of (R,R)-diisopropyl tartrate (0.148 g, 0.63 mmol) and an SmI_2-THF solution (0.1 M, 19 ml) was added dropwise a solution of ethylene glycol (0.5 ml) in THF (19 ml) during 30 min at room temperature. After being stirred for an additional 30 min, the reaction mixture was exposed to air to quench excess SmI_2. Ethylene glycol (0.57 ml), silica gel (*ca.* 3 g) and hexane (10 ml) were added, and the mixture was stirred for 10 min and filtered to remove silica gel. The silica gel was washed with hexane–ethyl acetate (3 : 1) and the filtrate was concentrated by a rotary evaporator. The residue was chromatographed on silica gel (hexane–ethyl acetate, 3 : 1) to afford (R)-diisopropyl malate (0.137 g, 99%) as an oil.

In relation to the reductions mentioned above, it has been reported that functionalized vinyloxiranes undergo facile reductive epoxide ring opening with SmI_2 to provide (E)-allylic alcohols [11]. The reactions are exceedingly rapid, taking place in most cases at –90 °C within minutes. Reactions transpire under essentially neutral conditions, so that little or no equilibration to form more stable olefinic isomers takes place. The unsaturated epoxide substrates are readily available from allylic alcohols via a three-step procedure. Therefore, this protocol provides an efficient synthetic route to optically active (E)-allylic alcohols.

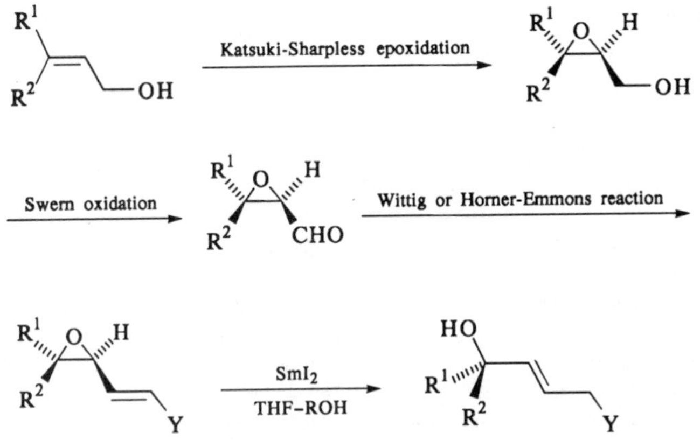

Y = CO$_2$Et, CH=CHCO$_2$Et, CN, COMe, COSEt, SO$_2$Ph, SPh, PO(OEt)$_2$, H, Me

Cyanohydrin diethyl phosphates, readily obtained from various ketones or aldehydes by reaction with diethyl phosphorocyanidate and lithium cyanide, react with SmI$_2$ to give the corresponding nitriles in excellent yields [12]. This method is applicable to α,β-unsaturated carbonyl compounds to give β,γ-unsaturated nitriles without isomerization of the double bonds. An example is shown below.

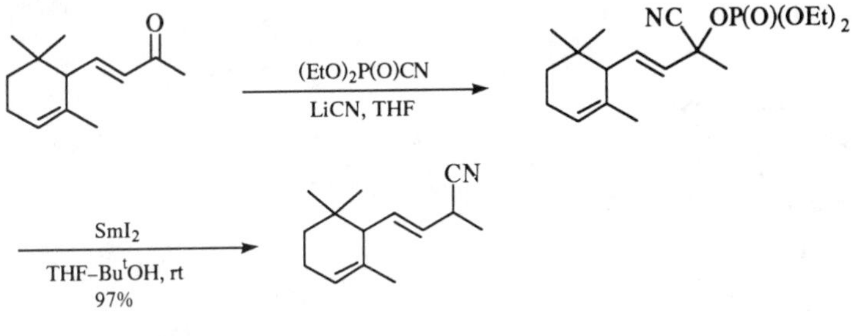

Preparation of a nitrile from a carbonyl compound. General procedure [12]

A carbonyl compound (0.5 mmol) was stirred with diethyl phosphorocyanidate (0.245 g, 1.5 mmol) and LiCN (0.050 g, 1.5 mmol) in THF (10 ml) for 10–30 min at room temperature. Water (10 ml) was added, and the mixture was extracted with ethyl acetate–hexane (1 : 1) (50 ml). The extract was washed with brine (2 x 20 ml), dried (MgSO$_4$), and evaporated under reduced pressure. A solution of the crude cyanophosphate and t-butanol (0.037 g, 0.5 mmol) in THF (5 ml) was added to a solution of SmI$_2$ prepared from samarium (0.345 g, 2.3 mmol) and diiodoethane (0.423 g, 1.5 mmol) in THF (10 ml). The mixture was stirred for an appropriate time with monitoring by TLC. The reaction mixture was quenched by the addition of 10%

HCl (10 ml) and extracted with ether (2 x 50 ml). The extracts were washed with 5% $Na_2S_2O_3$ (10 ml), H_2O (10 ml) and brine, and dried ($MgSO_4$). After removal of the solvent under reduced pressure, the residue was purified by column chromatography to give nitrile.

REFERENCES

1. G. A. Molander and G. Hahn, *J. Org. Chem.* **51,** 1135 (1986).
2. G. A. Molander and G. Hahn, *J. Org. Chem.* **51,** 2596 (1986).
3. K. Otsubo, J. Inanaga and M. Yamaguchi, *Tetrahedron Lett.* **28,** 4437 (1987).
4. K. Kusuda, J. Inanaga and M. Yamaguchi, *Tetrahedron Lett.* **30,** 2945 (1989).
5. J.-T. Hwang and C.-C. Liao, *Tetrahedron Lett.* **32,** 6583 (1991).
6. J. D. White and T. C. Somers, *J. Am. Chem. Soc.* **109,** 4424 (1987).
7. Y. Arai, M. Matsui and T. Koizumi, *J. Chem. Soc. Perkin Trans. 1,* 1233 (1990).
8. M. T. Reetz and E. H. Lauterback, *Tetrahedron Lett.* **32,** 4477 (1991).
9. S. Hanessian, C. Girard and J. L. Chiara, *Tetrahedron Lett.* **33,** 573 (1992).
10. J. Inanaga, J. Katsuki and M. Yamaguchi, *Chem. Lett.* 1025 (1991).
11. G. A. Molander, B. E. La Belle and G. Hahn, *J. Org. Chem.* **51,** 5259 (1986).
12. R. Yoneda, S. Harusawa and T. Kurihara, *J. Org. Chem.* **56,** 1827 (1991).

4.3.3 Carbonyl compounds

Carbonyl compounds are reduced by SmI_2 in the presence of a proton source to give alcohols [1]. The reaction is highly stereoselective and has been employed in total synthesis of natural products. An example for the synthesis of atractyligenin is illustrated below [2].

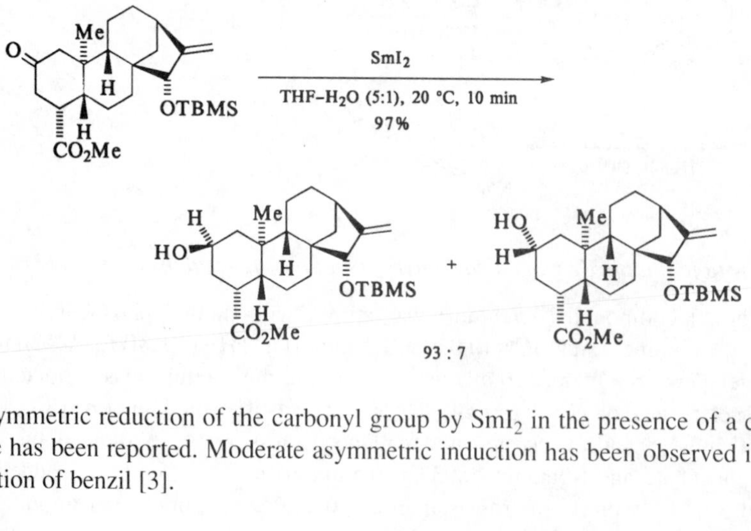

Asymmetric reduction of the carbonyl group by SmI_2 in the presence of a chiral amine has been reported. Moderate asymmetric induction has been observed in the reduction of benzil [3].

PhCOCOPh $\xrightarrow[\substack{\text{THF–HMPA, rt, 30 min} \\ 78\%}]{\text{SmI}_2\text{–quinidine}}$

$$\underset{\underset{\text{56% ee}}{}}{\text{Ph}\!-\!\overset{\overset{\displaystyle \text{HO}}{|}}{\underset{\underset{\displaystyle \text{O}}{\|}}{\text{C}}}\!-\!\overset{}{\underset{}{\text{C}}}\!-\!\text{Ph}}$$

   In the absence of a proton source, aldehydes and aromatic ketones are reductively coupled by SmI$_2$ to give pinacols (see section 4.4.2.1). Samarium(II) iodide is an efficient catalyst for the Meerwein–Ponndorf–Verley reduction of carbonyl compounds (see section 5.1.2).

REFERENCES

1.   P. Girard, J. L. Namy and H. B. Kagan, *J. Am. Chem. Soc.* **102**, 2693 (1980).
2.   A. K. Singh, R. K. Bakshi and E. J. Corey, *J. Am. Chem. Soc.* **109**, 6187 (1987).
3.   S. Takeuchi and Y. Ohgo, *Chem. Lett.* 403 (1988).

4.3.4 Carboxylic acids and their derivatives

Samarium(II) iodide is usually inert toward carboxylic acids, esters, amides and nitriles. However, these substrates are reduced by SmI$_2$ in strongly basic or acidic medium. Carboxylic acids or esters are very rapidly reduced by SmI$_2$ in the presence of NaOH, LiNH$_2$ or LiOMe to give primary alcohols [1–3]. Under similar conditions, amides are converted to primary alcohols as the major products [1], while in acidic medium (H$_3$PO$_4$) aldehydes are obtained in excellent yields [4]. Nitriles are reduced to primary amines in both basic and acidic media [4]. Some examples are shown in the following equations.

$$\text{C}_4\text{H}_9\text{CH(C}_2\text{H}_5)\text{CO}_2\text{H} \xrightarrow[\substack{\text{rt, 10 s} \\ 94\%}]{\text{SmI}_2, \text{ THF–H}_2\text{O-NaOH}} \text{C}_4\text{H}_9\text{CH(C}_2\text{H}_5)\text{CH}_2\text{OH}$$

$$\text{PhCONH}_2 \xrightarrow[\substack{\text{rt, 2 min} \\ 90\%}]{\text{SmI}_2, \text{ THF–H}_2\text{O-KOH}} \text{PhCH}_2\text{OH} \quad + \quad \text{PhCH}_2\text{NH}_2$$
$$91 : 9$$

$$p\text{-ClC}_6\text{H}_4\text{CONH}_2 \xrightarrow[\substack{\text{rt, 3 s} \\ 100\%}]{\text{SmI}_2, \text{ THF–85\% H}_3\text{PO}_4} p\text{-ClC}_6\text{H}_4\text{CHO}$$

$$\text{PhCN} \xrightarrow[\substack{\text{rt, 5 min} \\ 99\%}]{\text{SmI}_2, \text{ THF–85\% H}_3\text{PO}_4} \text{PhCH}_2\text{NH}_2$$

General procedure for the reduction of a carboxylic acid to a primary alcohol by
SmI$_2$–H$_3$PO$_4$ [4]

Under argon atmosphere, a solution of carboxylic acid (0.5 mmol) in THF (1 ml) was added to a solution of SmI$_2$ (0.1 M, 20 ml), and to this mixture was immediately added 85% H$_3$PO$_4$ (0.5 ml) by syringe. After the deep-blue colour of the reaction mixture changed to yellow-green, a chilled 25% NaOH solution (10 ml) was added and the mixture was stirred at room temperature for 10 min. The THF layer was separated and the aqueous phase was extracted with ether. The combined organic extracts were washed with water, then brine, and dried (MgSO$_4$). The solvent was evaporated and the residue was purified by chromatography on silica gel.

General procedure for the reduction of an amide to an aldehyde by
SmI$_2$–H$_3$PO$_4$ [4]

Under argon atmosphere, a solution of an amide (0.5 mmol) in THF (1 ml) was added to a solution of SmI$_2$ (0.1 M, 40 ml), and to this mixture was immediately added 85% H$_3$PO$_4$ (1 ml) by syringe, whereupon the typical blue colour of SmI$_2$ disappeared. After several minutes, 5% HCl (5 ml) was added to the resulting solution and stirred for 5 min. The mixture was extracted with ether and the combined extracts were washed with brine. The solvent was evaporated and the residue purified by chromatography on neutral alumina.

General procedure for the reduction of a nitrile to a primary amine by
SmI$_2$–H$_3$PO$_4$ [4]

Under argon atmosphere, a solution of a nitrile (1 mmol) in THF (1 ml) was added by syringe to a solution of SmI$_2$ (0.1 M, 40 ml), and 85% H$_3$PO$_4$ (1 ml) was immediately added by syringe. After the deep-blue colour of the reaction mixture changed to yellow-green, concentrated HCl (2 ml) was added to the resulting solution, and most of the solvent was distilled off under reduced pressure. The residue was diluted with water (2 ml) and the solution washed with ether. The aqueous phase was basified by the addition of 10% NaOH and was extracted with ether. After the usual work-up, the crude product was purified by chromatography on alumina to give the pure primary amine.

REFERENCES

1.   Y. Kamochi and T. Kudo, *Tetrahedron Lett.* **32,** 3511 (1991).
2.   Y. Kamochi and T. Kudo, *Chem. Lett.* 893 (1991).
3.   Y. Kamochi and T. Kudo, *Bull. Chem. Soc. Jpn* **65,** 3049 (1992).
4.   Y. Kamochi and T. Kudo, *Tetrahedron* **48,** 4301 (1992).

4.3.5 Miscellaneous substrates

Organoheteroatom oxides such as sulphoxides, sulphones, N-oxides, phosphine oxides, arsine oxides and tin oxides are reduced by SmI_2 [1–3]. The deoxygenations are accelerated by the addition of HMPA, and it is crucial for the reduction of sulphones, phosphine oxides or tin oxides [2].

Deoxygenation of epoxides to alkenes is effected by SmI_2 in THF–HMPA–DMAE, although the reaction is not stereospecific [1,4].

Alkenes are not reduced by SmI_2, while C–C double bonds conjugated with a carboxylic function are reduced at room temperature. Thus, α,β-unsaturated esters, acids and amides are subjected to double bond reduction by SmI_2 [1]. Addition of DMA or HMPA as a coqsolvent enhances the reduction rate [5,6].

Alkynes are also unreactive toward SmI_2 in THF, while in the presence of catalytic amounts of $CoCl_2$ or $FeCl_3$ they are reduced to alkenes. Either stereoisomer, (Z)- or (E)-alkene, can be prepared stereoselectively by changing the conditions [7] (Table 4.3).

$$Ph\!\!\equiv\!\!SiMe_3 \quad \xrightarrow[\text{cat. CoCl}_2 \cdot \text{4PPh}_3, \text{ 2 h}]{\text{SmI}_2} \quad PhCH\!=\!CHSiMe_3$$

TABLE 4.3

Solvent	Yield (%)	$Z:E$
MeOH	96	97 : 3
Pr^iOH–HMPA	92	6 : 94

Allylic acetates and propargylic acetates are subjected to reduction by SmI_2 in the presence or absence of Pd(0) catalyst [8–11]. The former compounds are usually converted to alkenes in moderate regio- and stereoselectivities. The latter ones lead to allenes or alkynes, allenes being the major products in most cases. This method, in particular the use of the SmI_2–Pd(PPh_3)_4$ reagent system, is useful in practical organic synthesis, as exemplified in the following schemes [12,13].

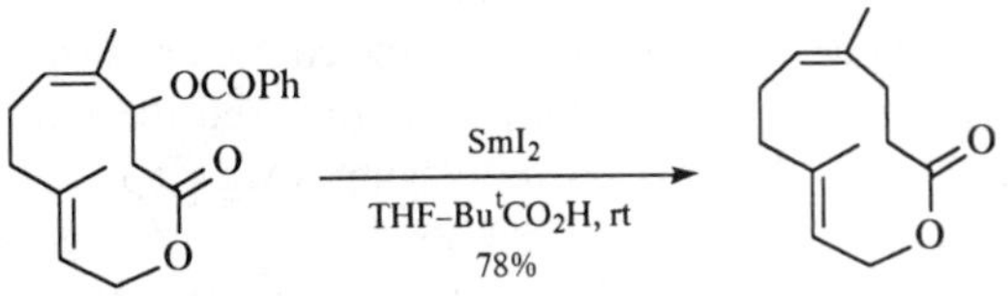

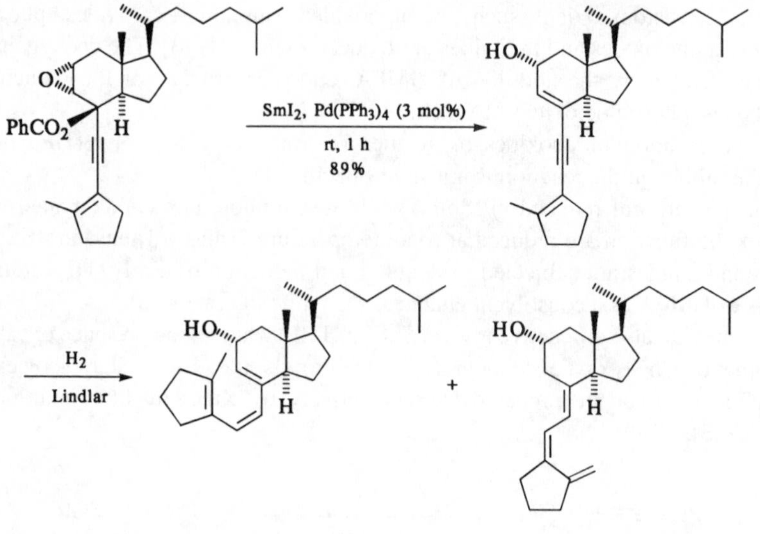

Reduction of (1S,2R,4S,6S)-2,4-diacetoxy-2-ethynyl-1-methyl-7-oxabicyclo[4.1.0]heptane to (3S,5R)-5-acetoxy-1-ethynyl-3-hydroxy-2-methylcyclohex-1-ene [12]

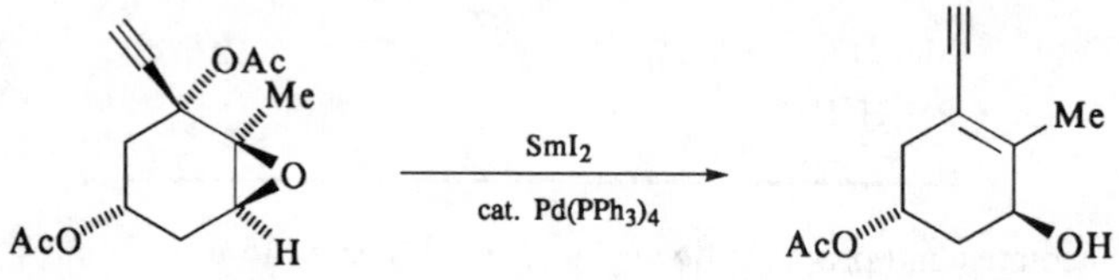

A solution of (1*S*, 2*R*, 4*S*, 6 *S*)-2,4-diacetoxy-2-ethynyl-1-methyl-7-oxabicyclo-[4.1.0]heptane (0.610 g, 2.42 mmol) and Pd(PPh$_3$)$_4$ (0.065 g, 0.056 mmol) in dry THF (28 ml) was added to a slurry of SmI$_2$ prepared from samarium (1.17 g, 7.78 mmol) and 1,2-diiodoethane (1.6 g, 5.67 mmol) in THF (21 ml). The mixture was stirred under argon at 25 °C for 4 h, after which time the blue colour persisted but none of the starting material remained (TLC). The mixture was poured into water (50 ml) and after it had been stirred for 5 min, the colour changed to brownish-green. Solid sodium carbonate was added until saturation and the mixture was extracted with ethyl acetate (3 x 50 ml). After drying (MgSO$_4$) and concentration of the solution, the residue was subjected to flash chromatography on silica gel (ethyl acetate–hexane, 1 : 3) to afford (3*S*,5*R*)-5-acetoxy-1-ethynyl-3-hydroxy-2-methylcyclohex-1-ene (0.430 g, 91%) as a light yellow oil: $[\alpha]_D^{25}$ −101.3° (*c* 3.0 in CHCl$_3$).

Arylsulphonyl groups are reductively removed by SmI_2. β-Hydroxysulphones and vicinal bissulphones are converted to alkenes [14–17]. Some alkyl sulphones are reduced to the corresponding alkanes under more forcing conditions [18]. A few representative examples are illustrated below.

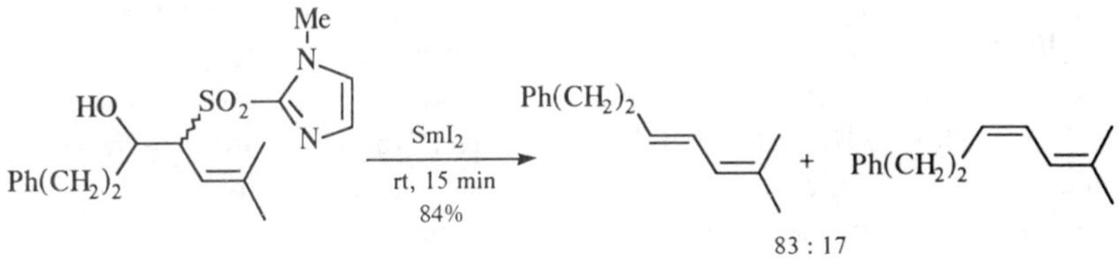

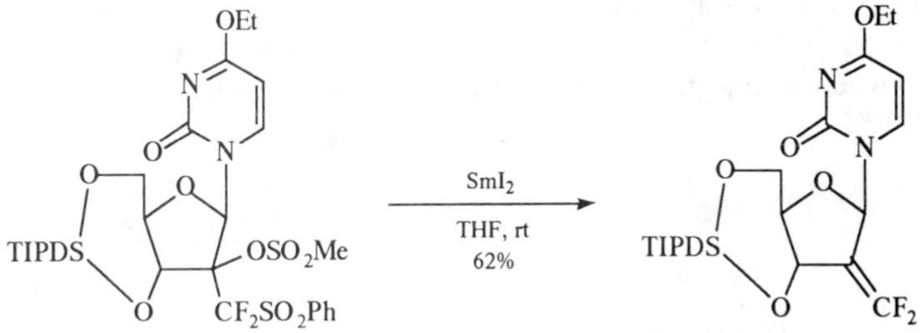

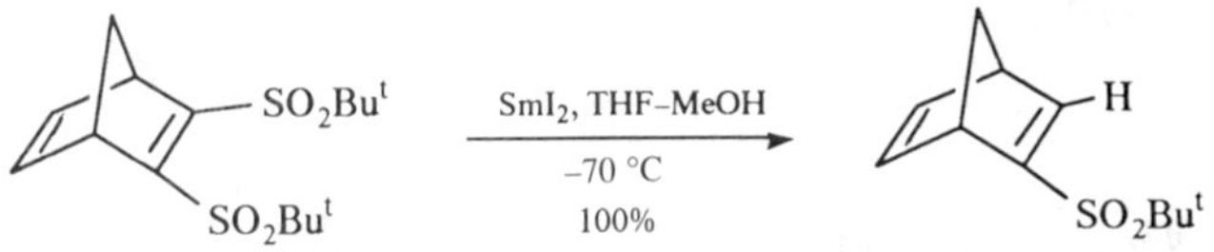

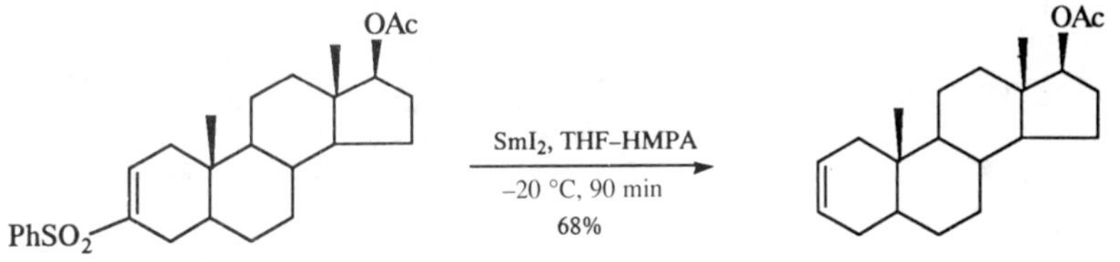

Reductive desulphonylation of an alkyl phenyl sulphone [18]

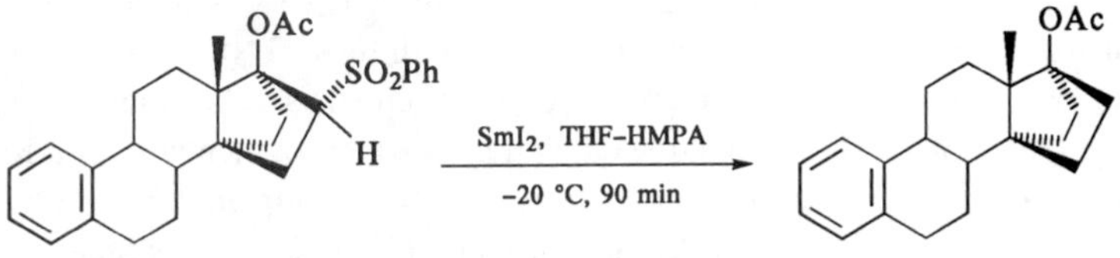

Sulphone **3** (0.494 g, 1 mmol) was dissolved in 50 ml of a freshly prepared SmI_2-THF (*ca.* 0.1 M) under argon. This mixture was cooled to –20 °C with stirring, and HMPA (CAUTION – CANCER SUSPECT AGENT) (4 ml) was added dropwise by syringe, whereupon the colour of the solution changed from blue to purple. After 90 min, the reaction was terminated with aqueous NH_4Cl (5 ml). Most of the THF was removed *in vacuo* by rotary evaporation. The product was precipitated by addition of cold HCl (0.5 ml) and isolated by suction-filtration. The solid residue was dissolved in ethyl acetate, and the resulting organic phase washed with aqueous $Na_2S_2O_3$, followed by brine, and then dried over Na_2SO_4. Chromatography of the crude product on silica gel (hexane–ethyl acetate, 9 : 1) provided compound **4** (0.308 g, 87%), m.p. 119–120 °C (acetone–hexane).

Nitro compounds are reduced to primary amines on treatment with more than 6 molar equivalents of SmI_2 [3, 19, 20]. Use of 4 molar equivalents of SmI_2 with a short reaction time gives hydroxylamines as major products [20].

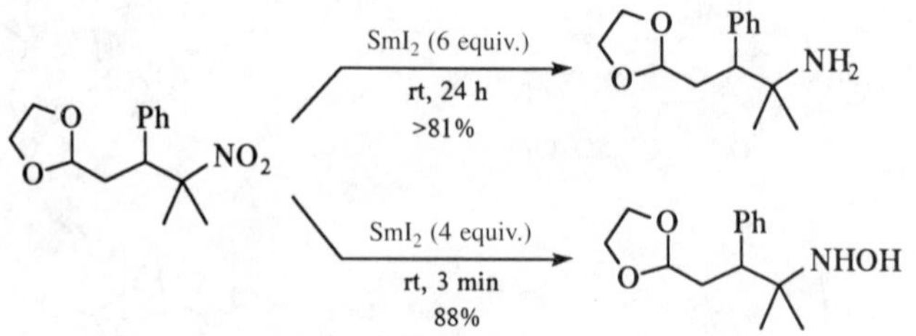

Reduction of a nitroalkane to a primary amine by SmI_2 [20]

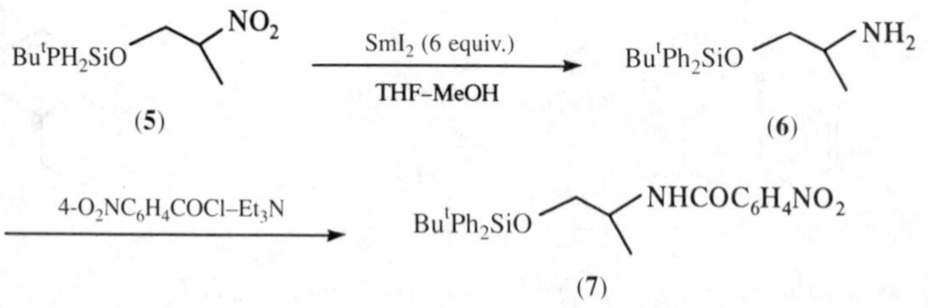

To a stirred solution of SmI_2 (4.2 mmol) in THF (30 ml) was added a solution of nitroalkane **5** (0.240 g, 0.7 mmol) in a 2 : 1 mixture of THF–MeOH (6 ml). The mixture was stirred for 8 h to give after work-up the crude amine **6**. This was dissolved in CH_2Cl_2 (15 ml) and treated with Et_3N (0.5 ml), then with 4-nitrobenzoyl chloride (0.427 g, 2.3 mmol), and stirred at room temperature overnight. The reaction mixture was diluted with ethyl acetate and water, the organic phase was separated, and the aqueous phase extracted with ethyl acetate several times. The combined extracts were dried and concentrated. The residue was chromatographed on silica gel (hexane–ethyl acetate, 85 : 15) to give compound **7** (0.256 g, 79%, m.p. 139–140 °C.

Relatively weak nitrogen–oxygen bonds of isooxazoles, isooxazolidines and oximes are cleanly cleaved by SmI$_2$ [21–23]. 2-Hydroxyimino amides of α-amino acid are reduced to optically active dipeptides with SmI$_2$ under mild conditions with high stereoselectivity [23].

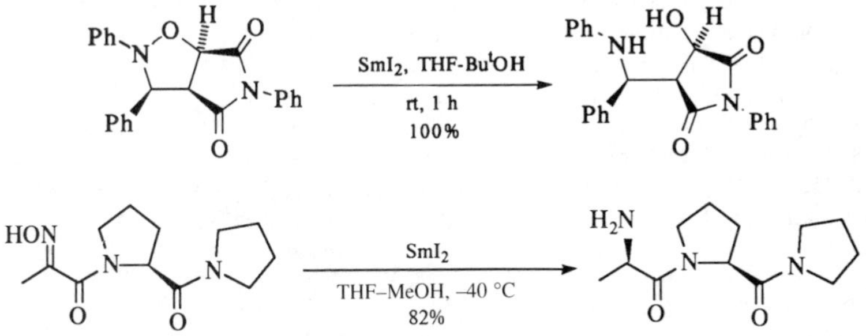

The N–N bond of *N*-benzoylhydrazines is cleaved almost instantaneously at 20 °C by SmI$_2$. This reaction is used for the preparation of optically active primary amines [24].

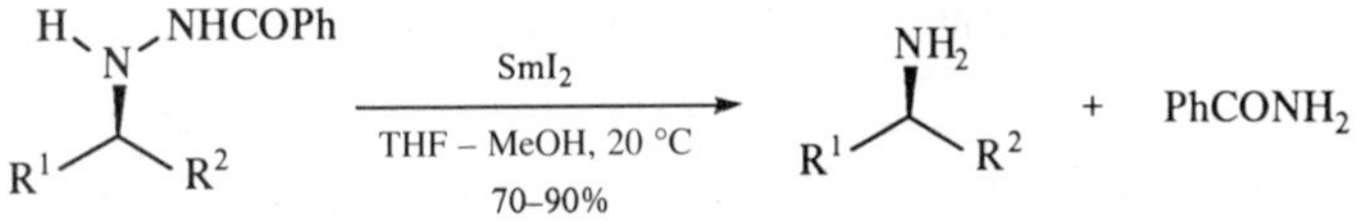

The oxygen–oxygen bond of cyclic peroxy compounds is readily cleaved by SmI$_2$ to give the corresponding diols. This reaction is utilized for the stereoselective synthesis of 1,3-diols in conjunction with the preparation of 1,2-dioxolanes from alkenylcyclopropanes [25].

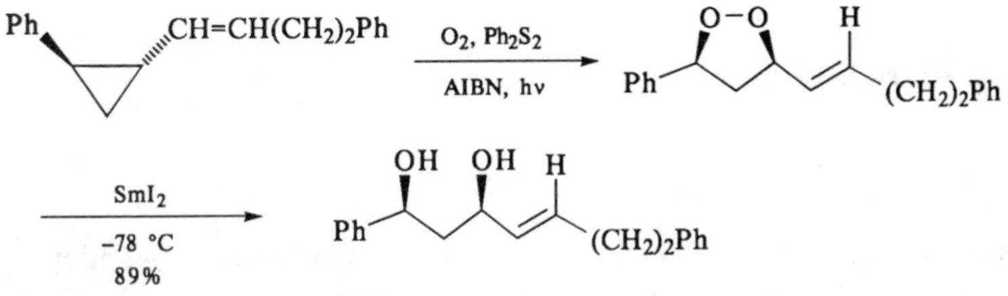

REFERENCES

1. P. Girard, J. L. Namy and H. B. Kagan, *J. Am. Chem. Soc.* **102**, 2693 (1980).
2. Y. Handa, J. Inanaga and M. Yamaguchi, *J. Chem. Soc. Chem. Commun.* 298 (1989).
3. Y. Zhang and R. Lin, *Synth. Commun.* **17**, 329 (1987).
4. M. Natsukawa, T. Tabuchi, J. Inanaga and M. Yamaguchi, *Chem. Lett.* 2101 (1987).
5. J. Inanaga, S. Sakai, Y. Handa, M. Yamaguchi and Y. Yokoyama, *Chem. Lett.* 2117 (1991).
7. A. Cabrera and H. Alper, *Tetrahedron Lett.* **33**, 5007 (1992).
8. J. Inanaga, Y. Yokoyama, Y. Baba and M. Yamaguchi, *Tetrahedron Lett.* **32**, 5559 (1991).
9. T. Tabuchi, J. Inanaga and M. Yamaguchi, *Tetrahedron Lett.* **27**, 601 (1986).
10. T. Moriya, Y. Handa, J. Inanaga and M. Yamaguchi, *Tetrahedron Lett.* **29**, 6947 (1988).
11. T. Tabuchi, J. Inanaga and M. Yamaguchi, *Tetrahedron Lett.* **27**, 5237 (1986).
12. W. H. Okamura, J. M. Aurrecoechea, R. A. Gibbs and A. W. Norman, *J. Org. Chem.* **54**, 4072 (1989)
13. J. D. Enas, G.-Y. Shen and W. H. Okamura, *J. Am. Chem. Soc.* **113**, 3873 (1991).
14. A. S. Kende and J. S. Mendoza, *Tetrahedron Lett.* **31**, 7105 (1990).
15. P. Pouilly, A. Chénedé, J.-M. Mallet and P. Sinay, *Tetrahedron Lett.* **33**, 8065 (1992).
16. J. S. Sabol and J. R. McCarthy, *Tetrahedron Lett.* **33**, 3101 (1992).
17. J. Belloch, M. Virgili, A. Moyano, M. A. Pericas and A. Riera, *Tetrahedron Lett.* **32**, 4579 (1991).
18. H. Künzer, M. Stahnke, G. Sauer and R. Wiechert, *Tetrahedron Lett.* **32**, 1949 (1991).
19. J. Souppe, L. Danon, J. L. Namy and H. B. Kagan, *J. Organometal. Chem.* **250**, 227 (1983).
20. A. S. Kende and J. S. Mendoza, *Tetrahedron Lett.* **32**, 1699 (1991).
21. N. R. Natale, *Tetrahedron Lett.* **23**, 5009 (1982).
22. T. Koizumi, private communication to the author of this book.
23. T. Mukaiyama, K. Yorozu, K. Kato and T. Yamada, *Chem. Lett.* 181 (1992).
24. M. J. Burk and J. E. Feaster, *J. Am. Chem. Soc.* **114**, 6266 (1992).
25. K. S. Feldman and R. E. Simpson, *Tetrahedron Lett.* **30**, 6985 (1989).

4.4 SAMARIUM(II) IODIDE-PROMOTED CARBON–CARBON BOND-FORMING REACTIONS

4.4.1 Carbonyl addition reactions

Although the preparation of organosamarium reagents such as Grignard reagent is not easy, Barbier-type and Reformatsky-type reactions can be accomplished by the use of SmI_2. These carbonyl addition reactions provide extremely valuable protocols which are hardly achievable by other methods. In particular, SmI_2-promoted intramolecular carbon–carbon bond-forming reactions have widespread synthetic applicability. This section describes many useful reactions which emphasize the characteristic properties of the divalent samarium.

4.4.1.1 Barbier-type reactions

4.4.1.1.1 Scope and intermolecular reactions

The reactions are largely dependent on substrates and reaction conditions. Simple alkyl iodides and aliphatic ketones undergo Barbier-type reaction in the presence of SmI_2 at reflux, although they require long reaction times [1,2]. Reactions of bromides are extremely sluggish and chlorides are virtually inert under these conditions. Addition of an iron(III) salt ($FeCl_3$ or $Fe(DBM)_3$) or $(C_5H_5)_2ZrCl_2$ as a catalyst significantly accelerates the reactions [1–3]. Addition of HMPA as a cosolvent dramatically accelerates the reactions [3]. For example, the coupling reaction of alkyl bromides with aliphatic ketones is complete within 1 min in THF–HMPA at room temperature.

$$C_4H_9I \quad + \quad C_6H_{13}COCH_3 \quad \xrightarrow[\substack{\text{reflux, 12 h} \\ 97\%}]{\text{2 SmI}_2,\ \text{THF}} \quad C_6H_{13}C(OH)(C_4H_9)CH_3$$

$$C_4H_9Br \quad + \quad C_6H_{13}COCH_3 \quad \xrightarrow[\substack{\text{reflux, 1.5 days} \\ 96\%}]{\text{2 SmI}_2,\ \text{THF}} \quad C_6H_{13}C(OH)(C_4H_9)CH_3$$

$$C_4H_9Br \quad + \quad C_6H_{13}COCH_3 \quad \xrightarrow[\substack{\text{rt, 1 min} \\ 92\%}]{\text{2 SmI}_2,\ \text{THF–HMPA}} \quad C_6H_{13}C(OH)(C_4H_9)CH_3$$

Unlike simple alkyl halides, activated halides such as benzyl bromide and allyl iodide undergo smooth SmI_2-promoted carbonyl addition at room temperature in THF without any additives [4,5]. Under similar conditions, benzyl chloromethyl ether [6], bromomethyl acetate [7], diiodomethane [8], bromomethyl p-methylphenyl sulphoxide [9] and carbon tetrachloride [9] react with carbonyl compounds. These reactions provide new protocols for nucleophilic introduction of one carbon unit possessing a functional group. Typical examples are illustrated in the following schemes. These protocols have potential synthetic utility in natural product synthesis, since the products are readily converted to 1,2-diols and epoxides.

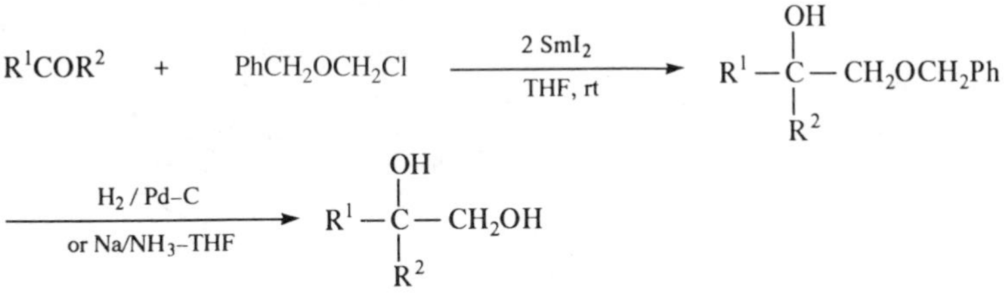

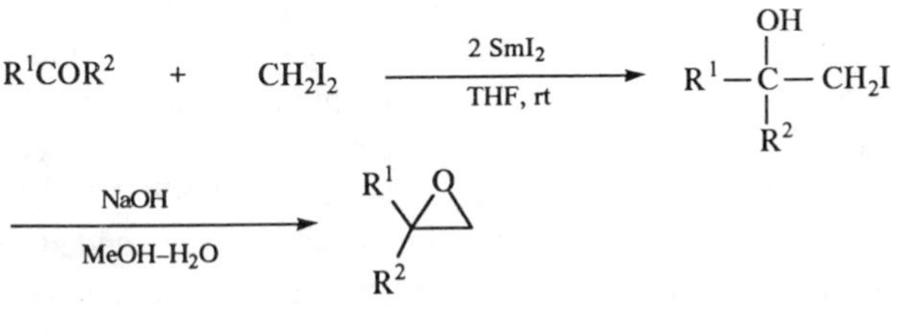

$$R^1COR^2 \quad + \quad PhCH_2OCH_2Cl \quad \xrightarrow[\text{THF, rt}]{\text{2 SmI}_2} \quad R^1 \!-\! \underset{\underset{R^2}{|}}{\overset{\overset{OH}{|}}{C}} \!-\! CH_2OCH_2Ph$$

$$\xrightarrow[\text{or Na/NH}_3\text{–THF}]{\text{H}_2\,/\,\text{Pd–C}} \quad R^1 \!-\! \underset{\underset{R^2}{|}}{\overset{\overset{OH}{|}}{C}} \!-\! CH_2OH$$

$$R^1COR^2 \quad + \quad CH_2I_2 \quad \xrightarrow[\text{THF, rt}]{\text{2 SmI}_2} \quad R^1 \!-\! \underset{\underset{R^2}{|}}{\overset{\overset{OH}{|}}{C}} \!-\! CH_2I$$

$$\xrightarrow[\text{MeOH–H}_2\text{O}]{\text{NaOH}} \quad \overset{R^1}{\underset{R^2}{\diagup}}\!\!\!\!\overset{O}{\triangle}$$

Analogous reactions take place between alkyl halides and isonitriles [10–12]. The initial formation of iminosamarium species proceeds essentially quantitatively and subsequent reaction with carbonyl compounds occurs in excellent yields. The adducts, α-hydroxy imines, are converted by acid-catalysed hydrolysis to the corresponding α-hydroxy ketones. This three-component coupling reaction can be applied to the synthesis of sugars and di- and tricarbonyl compounds, as exemplified in the following scheme [10].

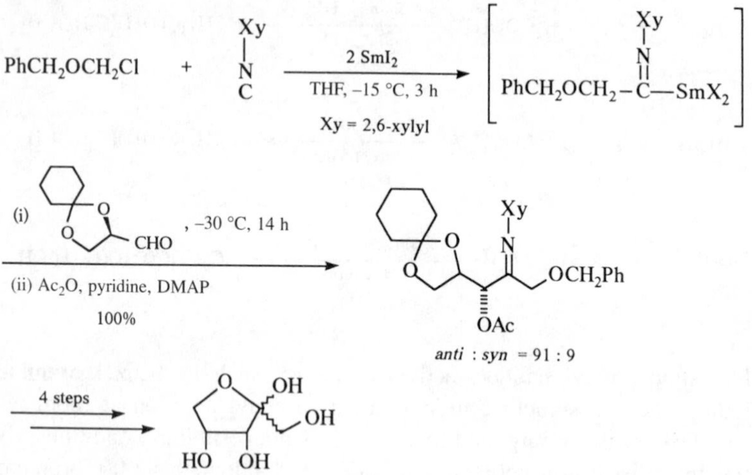

Samarium(II) iodide-promoted three-component coupling of bromoethane, 2,6-xylyl isonitrile and cyclohexanone; preparation of 1-[1-(2,6-xylylimino)-propyl]cyclohexanol [13]

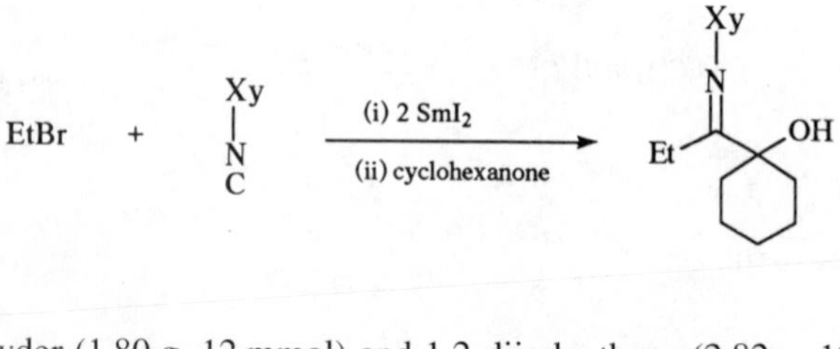

Samarium powder (1.80 g, 12 mmol) and 1,2-diiodoethane (2.82 g, 10 mmol) were placed in a round-bottomed flask containing a magnetic stirrer bar and a septum inlet under dry nitrogen atmosphere. The flask was immersed in an ice bath and THF (120 ml) was added with stirring. The ice bath was then removed and stirring continued for 1 h. HMPA (CAUTION – CANCER SUSPECT AGENT) (6 ml, 34 mmol) and 2,6-xylyl isocyanide (0.525 g, 4 mmol) were added and, after the reaction mixture had cooled to –15 °C, bromoethane (0.545 g, 5 mmol) was added.

The reaction mixture was stirred for 2 h at that temperature and then cyclohexanone (0.374 g, 3.81 mmol) was added. The reaction mixture was stirred for an additional 10 h, diluted with water (0.5 ml), ether (100 ml) and hexane (100 ml), and filtered through a short column of Florisil®. Column chromatography on silica gel (ether–hexane, 1 : 20) gave 1-[1-(2,6-xylylimino)propyl]cyclohexanol (0.934 g, 94%).

Allylic acetates react with carbonyl compounds by employing SmI_2 in the presence of Pd(0) complexes [14]. The reaction presumably proceeds via initial oxidative addition of Pd(0) to allylic acetates to generate η^3-palladium species, which subsequently react with SmI_2 to produce allylsamarium(III) and Pd(0) species. The allylsamarium(III) species react with carbonyl compounds to give addition products and the generated Pd(0) species again enter the catalytic cycle. The carbon–carbon bond formation tends to occur at the least substituted terminus of the allylic unit.

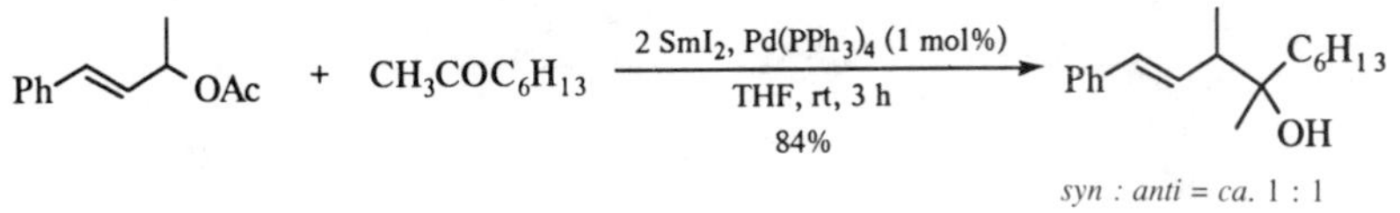

syn : anti = ca. 1 : 1

Propargyl acetates undergo an analogous reaction, in which allenic–propargylic samarium species are involved as reactive intermediates [15]. Selectivity of the formation of products, acetylenic and allenic alcohols, depends on substrates, and in most cases each product is produced with appreciable selectivity.

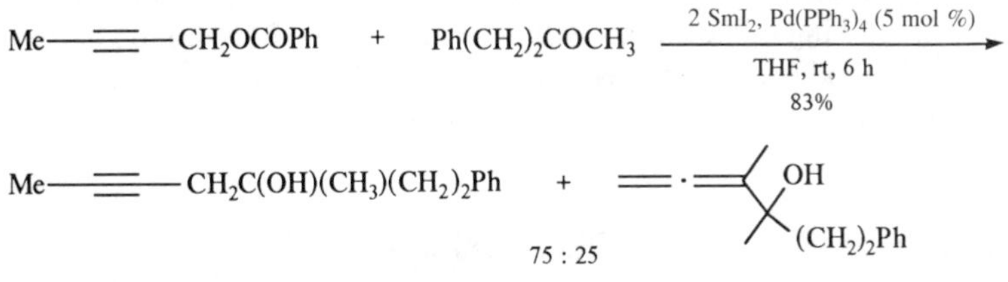

75 : 25

An interesting Barbier-type reaction of acid chlorides with ketones has been reported [16–18]. The cross-coupling reactions proceed very rapidly in THF or acetonitrile at room temperature to produce α-hydroxy ketones in moderate to high yields. Major by-products are homo-coupling products, α-diketones and pinacols. Use of a large excess of SmI_2 promotes deoxygenation of the hydroxy group [19].

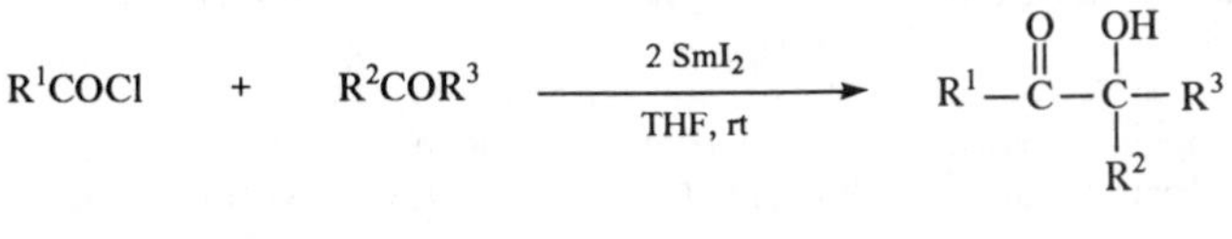

These cross-coupling reactions involve acylsamarium intermediates which are produced by successive two-electron transfers from SmI_2 to acid chlorides. A similar reaction occurs also using diphenylcarbamoyl chloride.

In contrast to these reactions, the reaction of α-alkoxy acid chlorides with ketones in the presence of SmI_2 produces 1,2-glycol monoethers [9,20]. Interestingly, one carbon degradation occurs in these reactions. The driving force of the decarbonylation is ascribed to the formation of a stabilized alkoxy radical. A similar cross-coupling reaction occurs between an N-protected amino acid chloride and a carbonyl compound [21].

Preparation of an α-hydroxy ketone from an acid chloride and a carbonyl compound. General procedure [17]

A mixture of an acid chloride (2 mmol) and a carbonyl compound (2 mmol) in THF (5 ml) was rapidly added to a solution of SmI_2 (0.1 M, 42 ml). The reaction mixture immediately turned green and was treated with 0.1 M HCl. Extraction with ether was followed by washing of the extract with water and brine. Evaporation of the solvent left the crude product, which was purified by flash chromatography on silica gel (hexane–ethyl acetate, 9 : 1) to give the α-hydroxy ketone.

REFERENCES

1. P. Girard, J. L. Namy and H. B. Kagan, *J. Am. Chem. Soc.* **102**, 2693 (1980).
2. H. B. Kagan, J. L. Namy and P. Girard, *Tetrahedron* **37**, Supplement 175 (1981).
3. K. Otsubo, K. Kawamura, J. Inanaga and M. Yamaguchi, *Chem. Lett.* 1487 (1987).
4. J. L. Souppe, J. L. Namy and H. B. Kagan, *Tetrahedron Lett.* **23**, 3497 (1982).
5. J. Souppe, L. Danon, J. L. Namy and H. B. Kagan, *J. Organometal. Chem.* **250**, 227 (1983).
6. T. Imamoto, T. Takeyama and M. Yokoyama, *Tetrahedron Lett.* **25**, 3225 (1984).
7. E. J. Enholm and H. Satici, *Tetrahedron Lett.* **32**, 2433 (1991).
8. T. Tabuchi, J. Inanaga and M. Yamaguchi, *Tetrahedron Lett.* **27**, 3891 (1986).
9. M. Sasaki, J. Collin and H. B. Kagan, *New J. Chem.* **16**, 89 (1992).
10. M. Murakami, T. Kawano and Y. Ito, *J. Am. Chem. Soc.* **112**, 2437 (1990).
11. M. Murakami, H. Masuda, T. Kawano, H. Nakamura and Y. Ito, *J. Org. Chem.* **56**, 1 (1991).
12. M. Murakami, T. Kawano, H. Ito and Y. Ito, *J. Org. Chem.* **58**, 1458 (1993).

13. An improved procedure on a larger scale. M. Murakami, T. Kawano, H. Ito and Y. Ito, private communication to the author.
14. T. Tabuchi, J. Inanaga and M. Yamaguchi, *Tetrahedron Lett.* **27**, 1195 (1986).
15. T. Tabuchi, J. Inanaga and M. Yamaguchi, *Chem. Lett.* 2275 (1987).
16. J. Souppe, J. L. Namy and H. B. Kagan, *Tetrahedron Lett.* **25**, 2869 (1984).
17. J. Collin, J. L. Namy, F. Dallemer and H. B. Kagan, *J. Org. Chem.* **56**, 3118 (1991).
18. S. M. Ruder, *Tetrahedron Lett.* **33**, 2621 (1992).
19. J. Collin, F. Dallemer, J. L. Namy and H. B. Kagan, *Tetrahedron Lett.* **30**, 7407 (1989).
20. H. B. Kagan, M. Sasaki and J. Collin, *Pure Appl. Chem.* **60**, 1725 (1988).
21. J. Collin, J. L. Namy, G. Jones and H. B. Kagan, *Tetrahedron Lett.* **33**, 2973 (1992).

4.4.1.1.2 Intramolecular reactions

Samarium(II) iodide-promoted intramolecular Barbier-type reactions are highly useful methods for the construction of carbocyclic compounds. These reactions have been extensively investigated by Molander and his coworkers, and the scope and limitations have been clarified [1–4]. Table 4.4 shows representative results.

Five- and six-membered rings can be constructed in excellent yields from 2-(ω-iodoalkyl)cycloalkanones. Excellent stereoselectivities are observed in the reactions of cyclopentanone substrates and 2-substituted-2-(ω-haloalkyl)cycloalkanones [1]. Similarly, bicyclo[$m.n.$1]alkan-1-ols are synthesized from a variety of 3-(ω-iodoalkyl)cycloalkanones [3]. The reaction is quite general; highly strained bicyclo[2.1.1]hexan-1-ol can be synthesized in good to high yields. Primary, secondary and tertiary iodides are all applicable and even sterically congested carbonyls are subjected to the reactions as well. This method, however, cannot be applied for the synthesis of [$m.n.$2]alkan-1-ols.

2-(ω-Iodoalkyl)-β-keto esters and 2-(ω-iodoalkyl)-β-keto amides undergo stereocontrolled cyclization by virtue of template effect of Sm^{3+} affording *cis*-2-hydroxycycloalkanecarboxylates and *cis*-hydroxycycloalkanecarboxamides, respectively [4].

Cyclization of ethyl 2,4-dimethyl-2-(3-iodopropyl)-3-oxopentanoate [4]

To samarium metal powder (0.30 g, 2.0 mmol), flamed and cooled under a slow flow of argon, was added THF (1 ml) followed by a solution of 1,2-diiodoethane (0.42 g, 1.5 mmol) in THF (2 ml). The mixture was stirred at ambient temperature for 1 h. The resulting solution was cooled to –78 °C in a dry ice–acetone bath, and ethyl 2,4-dimethyl-2-(3-iodopropyl)-3-oxopentanoate (0.255 g, 0.75 mmol) was slowly added. The reaction was allowed to stir and warm to room temperature overnight before the mixture was separated between saturated aqueous K_2CO_3 (5 ml) and ether (5 ml). The aqueous layer was extracted with ether (3 × 3 ml). The organic extracts were washed with brine (2 ml) and dried over $K_2CO_3/MgSO_4$. The solvent was evaporated and the residue distilled by the use of the Kugelrohr apparatus to give ethyl (1S*,2S*)-1-methyl-2-isopropyl-2-hydroxycyclopentanecarboxylate (0.140 g, 87%), b.p. 45–50 °C/0.05 mmHg.

β-Bromo ketones undergo SmI_2-promoted intramolecular Barbier-type reactions in the presence of HMPA to afford cyclopropanols. This reaction is utilized for the one-pot synthesis of cyclopropanol derivatives from ethyl β-bromopropionate and Grignard reagents [9].

Preparation of 1-butylcyclopropyl trimethylsilyl ether from ethyl
β-bromopropionate and butylmagnesium bromide [9]

$$Br(CH_2)_2CO_2Et \quad + \quad BuMgBr \quad \xrightarrow[\substack{\text{(ii) } H_2O \\ \text{(iii) TMSCl–pyridine}}]{\text{(i) } SmI_2, \text{ THF–HMPA}} \quad \overset{OTMS}{\underset{Bu}{\triangleright}}$$

Under a nitrogen atmosphere, ethyl β-bromopropionate (0.45 g, 2.5 mmol) was added to a THF–HMPA (CAUTION – CANCER SUSPECT AGENT) (10 ml, 1 ml) solution of SmI_2 (5 mmol) at −78 °C. After less than 10–15 s, n-butylmagnesium bromide (2 ml of 1 M ether solution) was injected by a syringe through a rubber septum over 1 min. The resulting mixture was stirred at −78 °C for 10 min and then the solution was allowed to react at room temperature for 30 min, during which time the deep-green colour of the SmI_2 solution became yellowish green. After the usual work-up, the mixture was extracted with ether (2 × 25 ml) and the combined extracts were dried over $MgSO_4$. Evaporation of the solvent left a yellow oil which was treated with TMSCl (0.20 g, 2 mmol) in pyridine (0.25 g, 3 mmol) at 25 °C for 2 h. The mixture was subjected to short-column chromatography on silica gel (hexane–ether, 5 : 1) to afford 1-butylcyclopropyl trimethylsilyl ether (0.34 g, 90%).

REFERENCES

1. G. A. Molander and J. Etter, *J. Org. Chem.* **51,** 1778 (1986).
2. G. A. Molander and J. B. Etter, *Synth. Commun.* **17,** 901 (1987).
3. G. A. Molander and J. A. McKie, *J. Org. Chem.* **56,** 4112 (1991).
4. G. A. Molander, J. B. Etter and P. W. Zinke, *J. Am. Chem. Soc.* **109,** 453 (1987).
5. H. Suginome and S. Yamada, *Tetrahedron Lett.* **28,** 3963 (1987).
6. G. Lannoye, K. Sambasivarao, S. Wehrli and J. M. Cook, *J. Org. Chem.* **53,** 2327 (1988).
7. J. J. Sosnowski, E. B Danaher and R. K. Murray, Jr., *J. Org. Chem.* **50,** 2759 (1985).
8. B. A. Barner, Y. Lin and M. A. Rahman, *Tetrahedron* **45,** 6101 (1989).
9. S. Fukuzawa, Y. Niimoto and S. Sakai, *Tetrahedron Lett.* **32,** 7691 (1991).

TABLE 4.4

Intramolecular Barbier-type Reactions Promoted by SmI_2

Substrate	Reaction conditions	Product(s) Yield %	Ref.
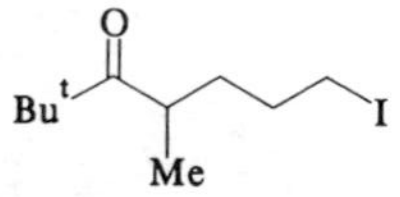	THF, −20 °C to 0 °C, <10 min	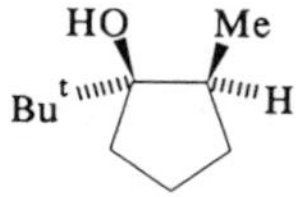 (92)	[2]
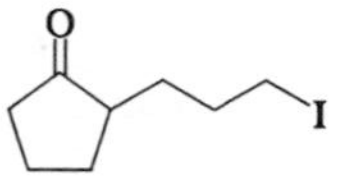	THF, Fe(DBM)$_3$ (cat.), rt	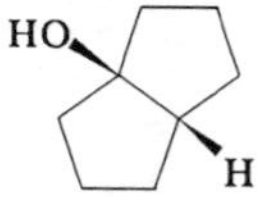 (90) 93% de	[1]
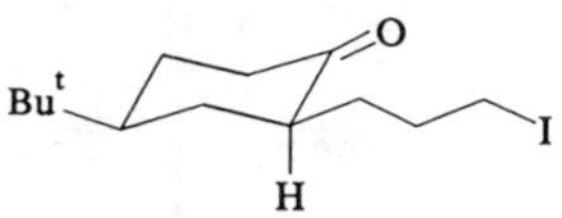	THF, Fe(DBM)$_3$ (cat.), rt	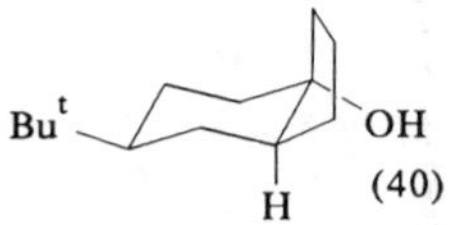 (40) 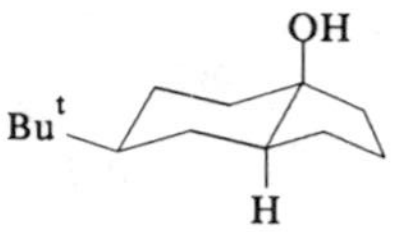 + (40)	[1]
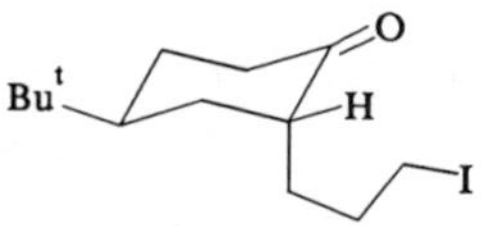	THF, Fe(DBM)$_3$ (cat.), rt	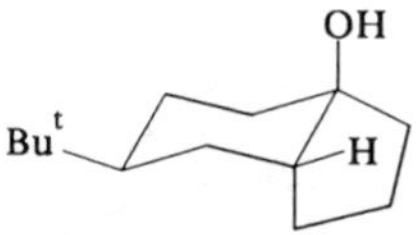 (80)	[1]

Continued

TABLE 4.4 *continued*

Intramolecular Barbier-type Reactions Promoted by SmI_2

Substrate	Reaction conditions	Product(s) Yield %	Ref.
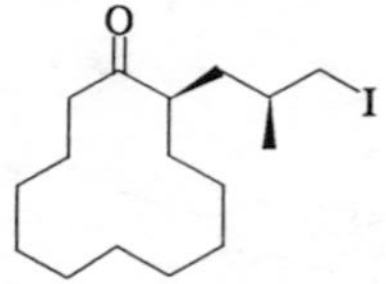	THF–HMPA, rt	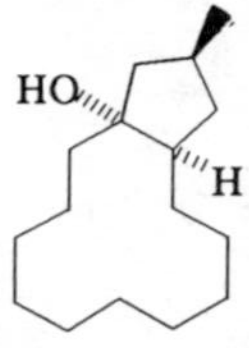 (90) 100% de	[5]
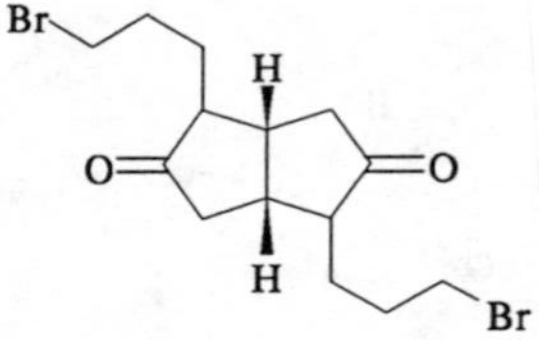	THF–(HMPA), FeCl$_3$ (cat.), high dilution conditions, rt, 2 days	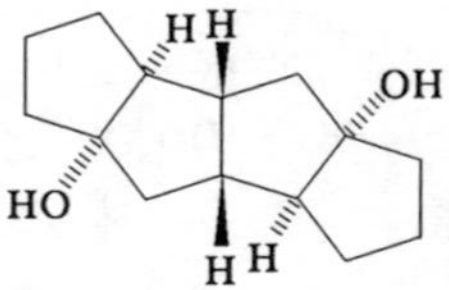 (80)	[6]
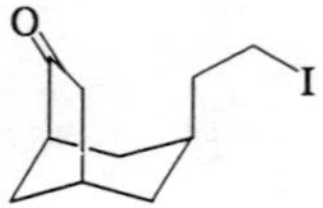	THF, FeCl$_3$ (cat.), rt, 15 h	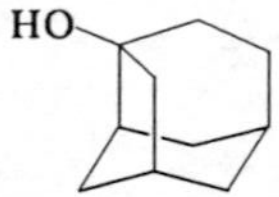 (44)	[7]
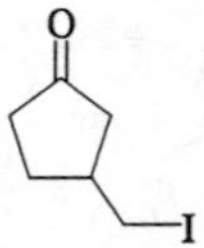	THF, Fe(DPM)$_3$ (cat.), −78 °C to 0 °C, 2 h	 (66)	[3]
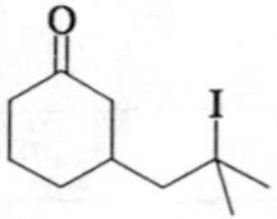	THF, Fe(DPM)$_3$ (cat.), −78 °C to 0 °C, 2h	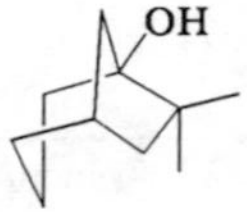 (57)	[3]

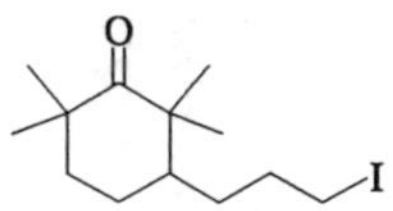

THF Fe(DPM)₃ (cat.), −78 °C to 0 °C, 2 h [3]

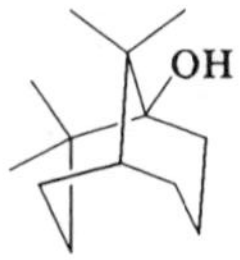

(74)

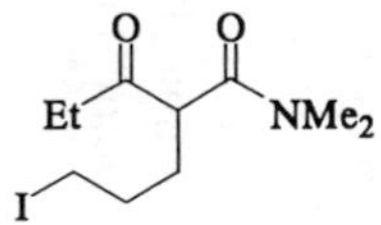

THF, −78 °C to rt [4]

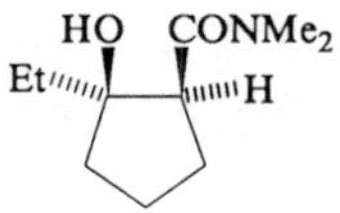

(79)

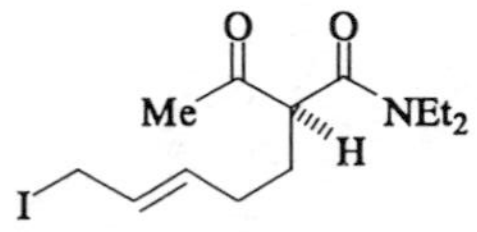

THF, −78 °C [4]

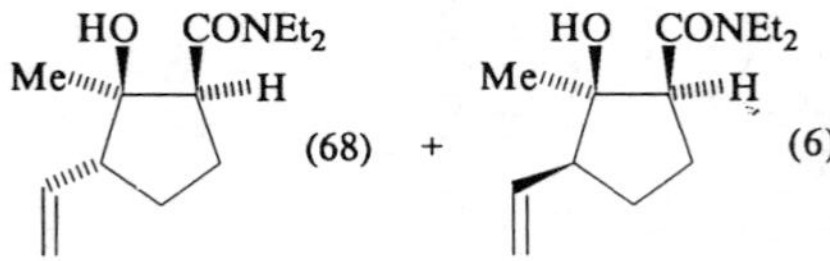

(6) + (68)

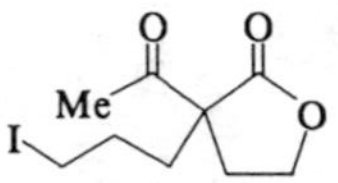

THF Fe(DPM)₃ (cat.), −78 °C to 0 °C, 2 h [4]

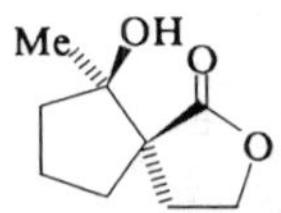

(100)

92% de

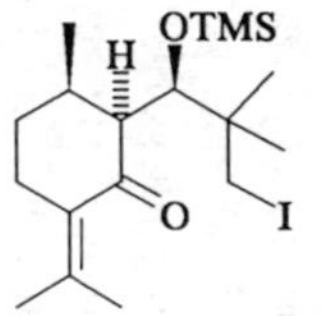

THF, rt, 3 h [8]

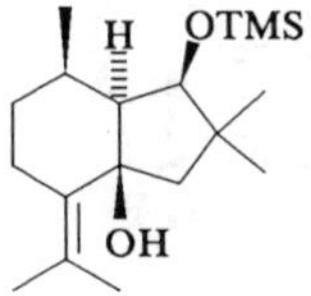

(96)

4.4.1.2 Reformatsky-type reactions

α-Bromo esters react with carbonyl compounds in the presence of SmI_2 to give β-hydroxy esters [1]. The reactions occur smoothly at room temperature, usually without any additives or catalysts. The related reactions using α-benzoate esters also occur in THF–HMPA [2].

The synthetic utility of SmI_2-promoted Reformatsky-type reactions has been proved in intramolecular reactions. For example, medium- and large-ring lactones and carbocycles can be synthesized in high yields by cyclizations of ω-(α-bromoacyloxy) aldehydes and α-bromo-ω-oxo esters [3,4]. One of the characteristic features of this method is that the reaction does not require high dilution conditions. This is reasonably explained by considering the six-membered transition state where the oxophilic samarium(III) plays a key role by interaction with the aldehyde oxygen atom. This protocol is especially useful for the synthesis of natural products [5,6].

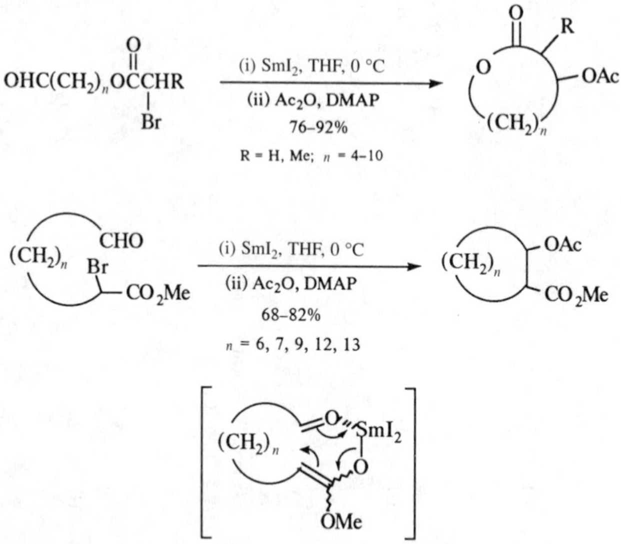

Asymmetric induction in the SmI_2-promoted intramolecular Reformatsky-type reaction has been extensively studied by Molander and his coworkers [7, 8]. In general exceedingly high (>200 : 1) 1,2- and 1,3-asymmetric inductions are observed in the cyclization of β-haloacetoxy carbonyl substrates leading to β-hydroxy-δ-lactones. It is noted that the 1,3-asymmetric induction is strictly controlled regardless of the substituent patterns at the α-position. These excellent stereoselectivities are responsible for the reactions proceeding through a rigid transition state enforced by chelation.

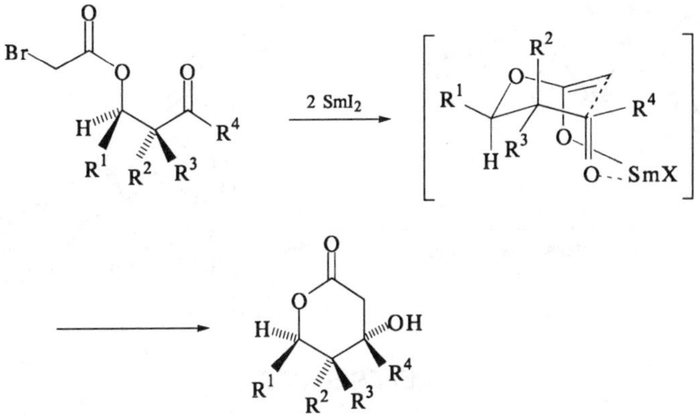

In the reactions of γ-haloacetoxy carbonyl substrate giving β-hydroxy ε-lactones, excellent 1,2-asymmetric induction is observed, while they result in moderate and low levels of 1,3- and 1,4-asymmetric inductions, respectively [8].

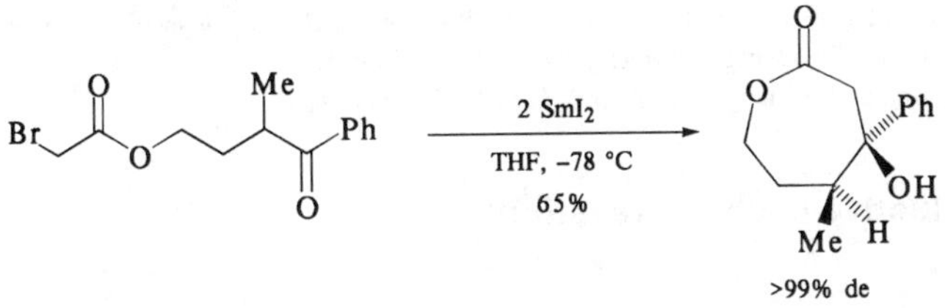

Intramolecular Reformatsky-type reaction. General procedure [8]

To a slurry of samarium metal powder (0.33 g, 2.1 mmol) in THF (15 ml) at 0 °C was added diiodomethane (0.536 g, 2 mmol). The mixture was stirred at 0 °C for 15 min and allowed to warm to room temperature for 1 h. The substrates were added at the appropriate temperature (usually –78 °C) and the mixture was stirred for approximately 1 h before being partitioned between saturated aqueous NH$_4$Cl (15 ml) and ether (15 ml). The aqueous layer was extracted with ether (3 x 5 ml), and the combined organic extracts were washed with brine (5 ml), dried over MgSO$_4$, filtered and concentrated. The product was isolated by flash chromatography on silica gel or by recrystallization.

A modified Reformatsky-type reaction of α-benzoate esters with ketones is promoted by SmI$_2$ in THF–HMPA [9,10]. The reactions with terpene ketones such as (–)-menthone and (+)-dihydrocarvone provide the corresponding β-hydroxy esters with excellent diastereoselectivity. An example is shown below [10].

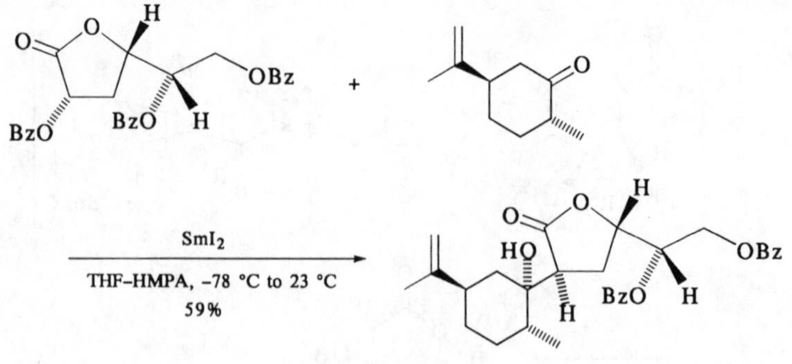

REFERENCES

1. H. B. Kagan, J. L. Namy and P. Girard, *Tetrahedron* **37,** Supplement 175 (1981).
2. E. J. Enholm and S. Jiang, *Tetrahedron Lett.* **33,** 313 (1992).
3. T. Tabuchi, K. Kawamura, J. Inanaga and M. Yamaguchi, *Tetrahedron Lett.* **27,** 3889 (1986).
4. E. Vedejs and S. Ahmad, *Tetrahedron Lett.* **29,** 2291 (1988).
5. T. Moriya, Y. Handa, J. Inanaga and M. Yamaguchi, *Tetrahedron Lett.* **29,** 6947 (1988).
6. J. Inanaga, Y. Yokoyama, Y. Handa and M. Yamaguchi, *Tetrahedron Lett.* **32,** 6371 (1991).
7. G. Molander and J. B. Etter, *J. Am. Chem. Soc.* **109,** 6556 (1987).
8. G. A. Molander, J. B. Etter, L. S. Harring and P.-J. Thorel, *J. Am. Chem. Soc.* **113,** 8036 (1991).
9. E. J. Enholm and S. Jiang, *Tetrahedron Lett.* **33,** 313 (1992).
10. E. J. Enholm and S. Jiang, *Tetrahedron Lett.* **33,** 6069 (1992).

4.4.2 Reductive coupling reactions

4.4.2.1 Pinacol coupling and related reactions

Carbonyl compounds, on treatment with SmI_2 in the absence of a proton source, undergo reductive dimerization to pinacols [1,2]. The reactions are usually complete within a few minutes at room temperature for aromatic aldehydes and ketones; a few hours are needed for aliphatic aldehydes, and a day is needed for aliphatic ketones. In the reactions of aromatic carbonyl compounds, substituents such as CN, NO_2 and CO_2R remain intact if a stoichiometric amount of SmI_2 is used. Intermolecular cross-coupling is possible between aldehydes and pyruvic acid esters of α-diketones to produce α,β-dihydroxycarbonyl substrates in satisfactory yields [3,4]. Intermolecular pinacol couplings by SmI_2 usually result in the formation of the mixture of ($\pm$) and mesostereoisomers with low stereoselectivity.

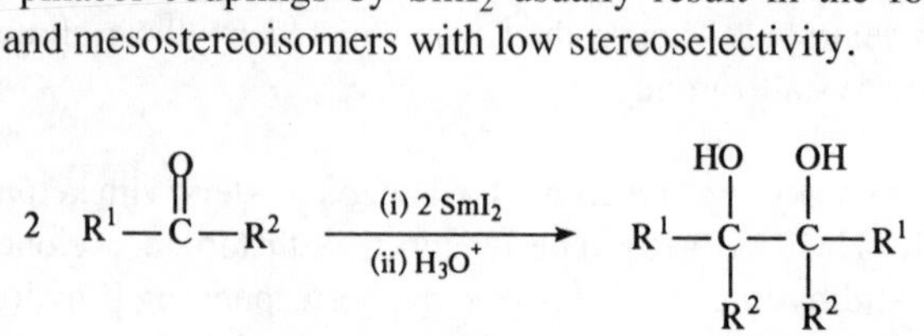

Intramolecular pinacol coupling promoted by SmI_2 is versatile in the construction of cyclic vicinal diols [5–8]. In most cases, *cis*-diols are formed in good to high selectivity, and in some cases they are produced exclusively. These processes provide stereochemically complementary carbocycles and therefore may have great synthetic value.

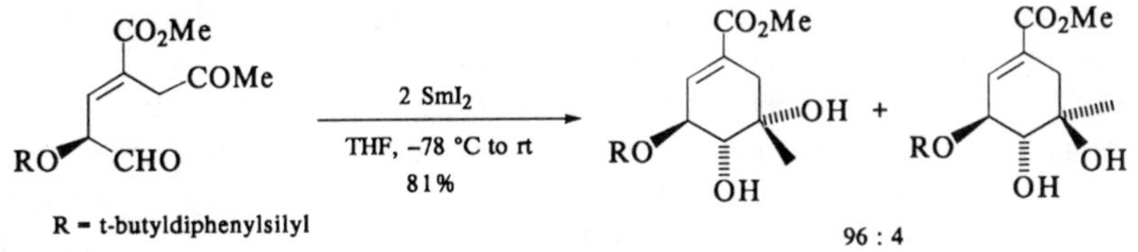

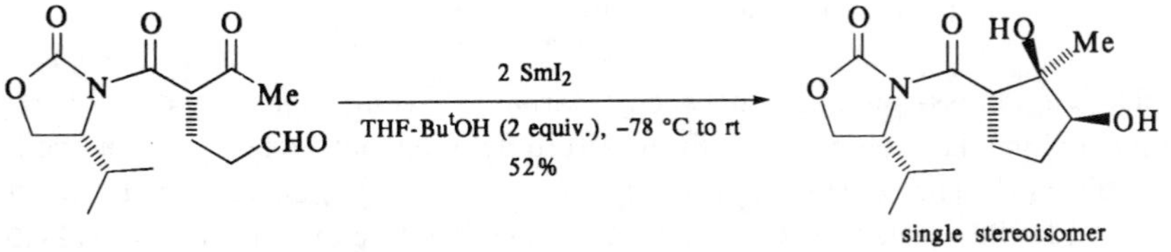

R = t-butyldiphenylsilyl

96 : 4

single stereoisomer

Reductive cyclization of ethyl 2-acetyl-2-methyl-5-oxopentanoate [6]

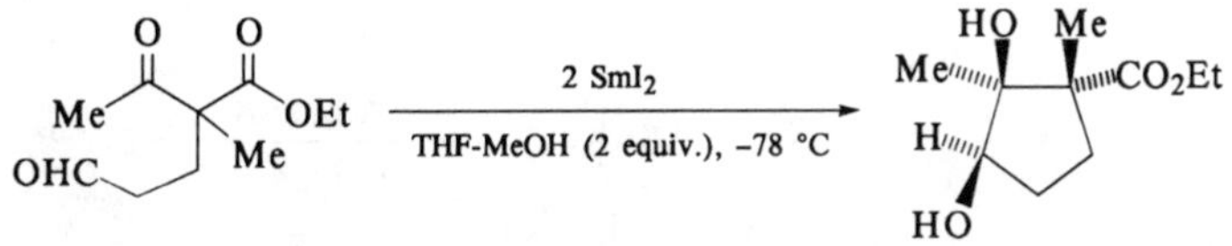

A slurry of SmI_2 (2 mmol) in THF (10 ml) was cooled to –78 °C, and a solution of ethyl 2-acetyl-2-methyl-5-oxopentanoate (0.200 g, 1 mmol) and MeOH (2 mmol) in dry THF (10 ml) was added dropwise. The resulting solution was stirred at –78 °C for 1–2 h and then slowly warmed to room temperature before quenching with saturated $NaHCO_3$. The layers were separated and the aqueous layer was extracted with ethyl acetate. The combined organic extracts were washed with saturated NaCl, dried (Na_2SO_4), filtered and concentrated *in vacuo* to provide ethyl (1R*,2S*,3R*)-2,3-dihydroxy-1,2-dimethylcyclopentanecarboxylate as a 200 : 1 mixture of diastereomers. The product was isolated as a clear colourless liquid (0.154 g, 77%) by flash chromatography on silica gel (hexane–ethyl acetate, 1 : 2).

Samarium(II) iodide is capable of reductive dimerization of aldimines to yield vicinal diamines [9,10]. This reaction requires more forcing conditions than the pinacol coupling reaction.

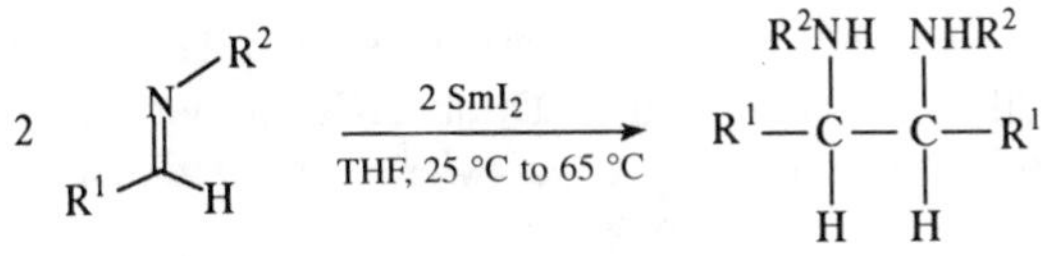

Reductive coupling of an imine to 1,2-diamines. General procedure [9]

A 0.1 M solution of SmI_2 in THF (3.0 equiv.) was added to a dry, round-bottomed flask equipped with a stir bar and a reflux condenser under an inert atmosphere of argon. The aldimine (1.0 equiv.) was added to the solution in THF by syringe and heated slowly to reflux. Occasionally, an additional equivalent of SmI_2 was added to maintain the blue-green colour throughout the heating process. The mixture was heated for periods of time ranging from 6 h to 12 h while monitored by TLC. When the aldimine was consumed, the reaction vessel was allowed to cool and MeOH–silica gel was added and stirred. After 0.5 h, the mixture was suction filtered through silica gel to remove most of the samarium salts. The silica gel was washed with ethyl acetate and CH_2Cl_2. Removal of the solvents on the rotary evaporator produced the crude 1,2-diamine product as a dark-brown oil. The pure diamine was obtained via flash column chromatography.

Reductive cross-coupling between carbonyl compounds and *O*-benzyl formaldoxime is promoted by SmI_2 affording the corresponding aminomethyl alcohols [11]. The *N*-benzyloxy group of the coupling products can be readily removed by PtO_2-catalysed hydrogenolysis, and therefore this method is a valuable entry for the synthesis of 1,2-amino alcohols. The reactions are postulated to proceed through the initial ketyl formation and subsequent addition to the carbon–nitrogen double bond.

$$R^1COR^2 \quad + \quad CH_2{=}NOCH_2Ph \quad \xrightarrow[\text{THF–HMPA, rt}]{\text{2 SmI}_2\text{, ROH}}$$

$$R^1 - \underset{\underset{R^2}{|}}{\overset{\overset{OH}{|}}{C}} - CH_2NHOCH_2Ph \quad \xrightarrow{H_2 / PtO_2} \quad R^1 - \underset{\underset{R^2}{|}}{\overset{\overset{OH}{|}}{C}} - CH_2NH_2$$

Aliphatic and aromatic acid chlorides react with 2 equivalents of SmI_2 to give α-diketones in good to high yields [12]. These coupling reactions involve acyl anion intermediates.

$$2\,RCOCl \quad \xrightarrow[\text{THF, rt, <20 min}]{\text{2 SmI}_2} \quad RCOCOR$$

Reductive coupling of benzoyl chloride to benzil [12]

A solution of benzoyl chloride (0.351 g, 2.5 mmol) in THF (2 ml) was slowly added under nitrogen at room temperature to a 0.1 M SmI_2 THF solution (50 ml). The mixture turned yellow in a few minutes. Dilute HCl was added, and the mixture was extracted with ether. The combined extracts were washed with aqueous $NaHCO_3$ and dried over $MgSO_4$. The solvent was evaporated and the residue recrystallized to give benzil (0.205 g, 78%), m.p. 92–94 °C.

Amides are not reduced by SmI_2 under ordinary conditions (THF, reflux). However, the coupled use of Sm and SmI_2 promotes reductive coupling of amides to *vic*-diaminoalkenes [13]. The actual reducing agent is supposed to be SmI which is generated by the reaction of SmI_2 with metallic samarium.

$$2\ ArCONR_2 \xrightarrow[\text{THF, reflux}]{\text{Sm–SmI}_2}$$

REFERENCES

1. J. L. Namy, J. Souppe and H. B. Kagan, *Tetrahedron Lett.* **24**, 765 (1983).
2. J. Souppe, L. Danon, J. L. Namy and H. B. Kagan, *J. Organometal. Chem.* **250**, 227 (1983).
3. K. Abe, S. Takahashi and N. Mori, *Chem. Express* **6**, 53 (1991).
4. N. Miyoshi, S. Takeuchi and Y. Ohgo, *Chem. Lett.* 959 (1993).
5. G. A. Molander and C. Kenny, *J. Org. Chem.* **53**, 2132 (1988).
6. G. A. Molander and C. Kenny, *J. Am. Chem. Soc.* **111**, 8236 (1989).
7. J. L. Chiara, W. Cabri and S. Hanessian, *Tetrahedron Lett.* **32**, 1125 (1991).
8. J. Uenishi, S Masuda and S. Wakabayashi, *Tetrahedron Lett.* **32**, 5097 (1991).
9. E. J. Enholm, D. C. Forbes and D. P. Holub, *Synth. Commun.* **20**, 981 (1990).
10. T. Imamoto and S. Nishimura, *Chem. Lett.* 1141 (1990).
11. T. Hanamoto and J. Inanaga, *Tetrahedron Lett.* **32**, 3555 (1991).
12. P. Girard, R. Couffignal and H. B. Kagan, *Tetrahedron Lett.* **22**, 3959 (1981).
13. A. Ogawa, N. Takami, M. Sekiguchi, I. Ryu, N. Kambe and N. Sonoda, *J. Am. Chem. Soc.* **114**, 8729 (1992).

4.4.2.2 Carbonyl unsaturated bond coupling

4.4.2.2.1 Intermolecular reactions

Ketyl radicals generated by the reaction of carbonyl compounds with SmI_2 add to alkenes to form carbon–carbon bonds [1–4]. These reactions have a broad scope for a variety of combinations of carbonyl and alkene substrates. Various aldehydes and ketones including aromatic ones can be employed. A variety of substituted olefins are applicable, in that terminal olefins and conjugated olefins such as α,β-unsaturated esters undergo smooth coupling reactions. The presence of a proton donor is essential for obtaining satisfactory results, otherwise many by-products are created. The reactions are usually accelerated in the presence of HMPA. The intermolecular coupling reactions of carbonyl compounds with α,β-unsaturated esters permit subsequent cyclization to give γ-lactones.

$$R^1COR^2 \;+\; \underset{R^4}{\overset{R^3}{\diagdown}}\!\!=\!\!\underset{R^5}{\overset{CO_2R}{\diagup}} \quad\xrightarrow[\text{THF, ROH}]{2\ SmI_2}\quad \begin{array}{c} R^1 \\ R^2\!-\!\!\boxed{\ \ }\!\!-\!\!O \\ R^3\!-\!\qquad=\!O \\ R^4\ R^5\quad H \end{array}$$

Reactions of ethyl acrylate with a carbonyl compound in the presence of SmI₂;
preparation of a γ-lactone [2]

A THF (5 ml) solution of 1,2-diiodoethane (1.13 g, 4 mmol) or diiodomethane (1.07 g, 4 mmol) was added at room temperature to samarium metal powder (0.6 g, 4 mmol). The mixture was then stirred at ambient temperature until the colour of the solution became deep green (0.5–1 h). A mixture of ethyl acrylate (0.20 g, 2 mmol), a carbonyl compound (2 mmol) and t-butanol (0.14 g, 2 mmol) in THF (5 ml) was then added simultaneously at 0 °C (for aldehydes) or at room tempera-ture (for ketones). The resulting mixture was stirred for 2–10 h at the same tem-perature, during which time its colour changed to brownish yellow. The solution was poured into dilute HCl and the aqueous layer was extracted with ether (3 x 20 ml); the extract was dried (MgSO₄) and evaporated to leave a yellow oil which was subjected to flash column chromatography (hexane–ether, 1 : 1). The product was further purified by microdistillation.

The method is extended to carbonyl–alkyne systems. Aliphatic ketones and aldehydes can be reductively coupled with terminal alkynes in THF–HMPA at room temperature in the presence of t-butanol [5]. This protocol provides an efficient entry for the synthesis of allylic alcohols.

$$R^1COR^2 \;+\; R^3\!\!-\!\!\!\equiv\!\!\!-\!X \quad\xrightarrow[\text{Bu}^t\text{OH, rt, 5 min}]{2\ SmI_2,\ \text{THF–HMPA}}\quad \begin{array}{c} R^1 \quad OH \\ \diagup\;\diagdown \\ R^2 \qquad \diagdown\!\!\!=\!\!\!X \\ R^3 \end{array}$$

$$R^3 = H, Me, TMS$$
$$X = \text{alkyl, } CH_2OAc, Ph, TMS$$

REFERENCES

1.　S. Fukuzawa, A. Nakanishi, T. Fujinami and S. Sakai, *J. Chem. Soc. Chem. Commun.* **624** (1986).
2.　S. Fukuzawa, A. Nakanishi, T. Fujinami and S. Sakari, *J. Chem. Soc. Perkin Trans. 1,* **1669** (1988).
3.　K. Otsubo, J. Inanaga and M. Yamaguchi, *Tetrahedron Lett.* **27,** 5763 (1986).
4.　O. Ujikawa, J. Inanaga and M. Yamaguchi, *Tetrahedron Lett.* **30,** 2837 (1989).
5.　J. Inanaga, J. Katsuki, O. Ujikawa and M. Yamaguchi, *Tetrahedron Lett.* **32,** 4921 (1991).

4.4.2.2.2 Intramolecular reactions

Carbonyl compounds bearing an alkenyl or an alkynyl moiety at the appropriate position undergo intramolecular cyclization upon treatment with SmI₂ [1–5]. The reactions proceed best for the formation of five-membered ring systems. High diastereoselectivities are generally observed in the reaction of cyclopentanone or cyclohexanone derivatives to afford fused bicyclic ring systems [6–8].

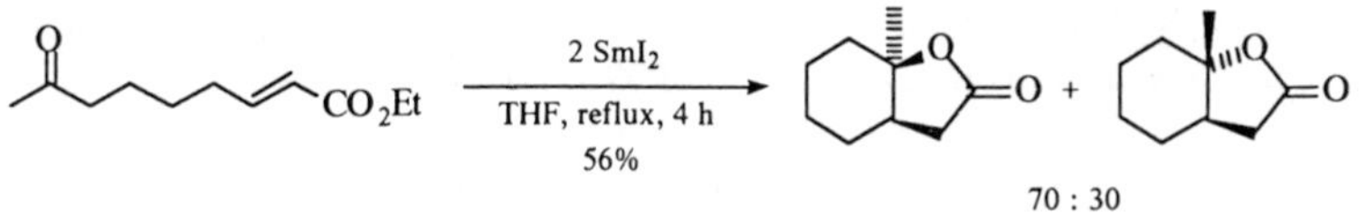

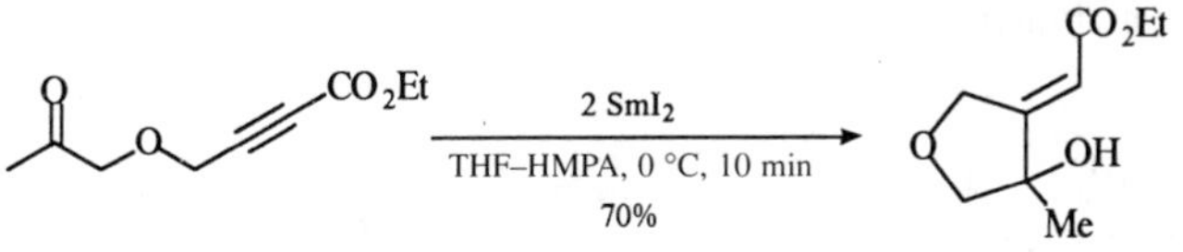

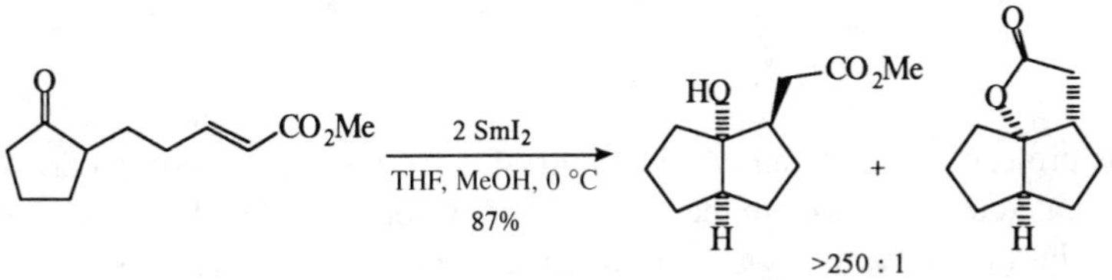

This method is applicable to unactivated olefinic ketones [9–12]. Organosamarium(III) species, generated by radical cyclization and subsequent reduction by SmI$_2$, are trapped by various electrophiles such as carbonyl compounds, disulphides, acid chlorides, oxygen and carbon dioxide [10].

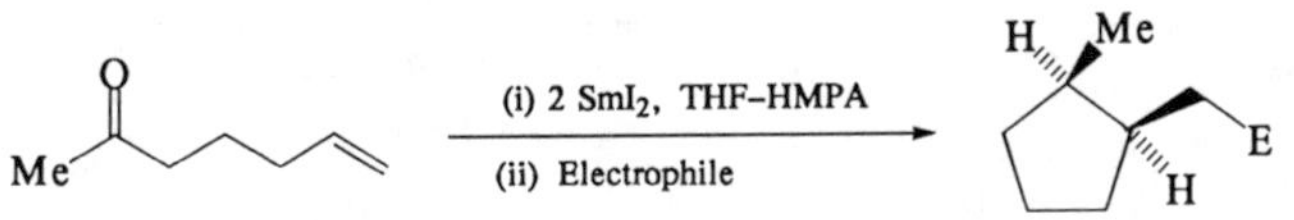

Reaction of 6-hepten-2-one with cyclohexanone in the presence of SmI$_2$; preparation of (1R*, 2S*)-1-[(2-hydroxy-2-methyl-1-cyclopentyl)methyl]-cyclohexan-1-ol [10]

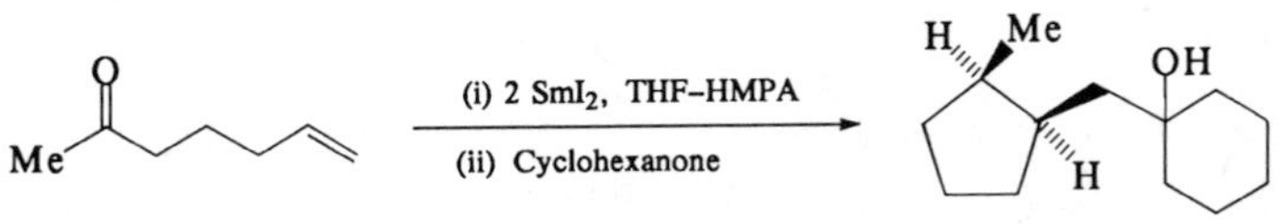

To a suspension of SmI$_2$ (2.25 mmol) in 16 ml of THF was added 3.6 ml (20.7 mmol) of HMPA (CAUTION – CANCER SUSPECT AGENT). The resulting solution was allowed to stir for 5 min, followed by the addition of 6-hepten-2-one (0.110 g, 0.98 mmol) in 20 ml of THF over 20 min. During this time a precipitate

(0.110 g, 0.98 mmol) in 20 ml of THF over 20 min. During this time a precipitate formed, which immediately went into solution upon addition of a solution of cyclohexanone (0.147 g, 0.150 mmol) in 3 ml of THF. After an additional 30 min the reaction was quenched by the addition of saturated aqueous $NaHCO_3$. The usual work-up and flash chromatography afforded $(1R^*,2S^*)$-1-[(2-hydroxy-2-methyl-1-cyclopentyl) methyl]-cyclohexan-1-ol (0.166 g, 80%), m.p. 104–105 °C.

The reactions of derivatives of 3-keto esters or amides also provide excellent diastereoselectivities, where the generated Sm^{3+} ion enforces the formation of a rigidly chelated intermediate [11].

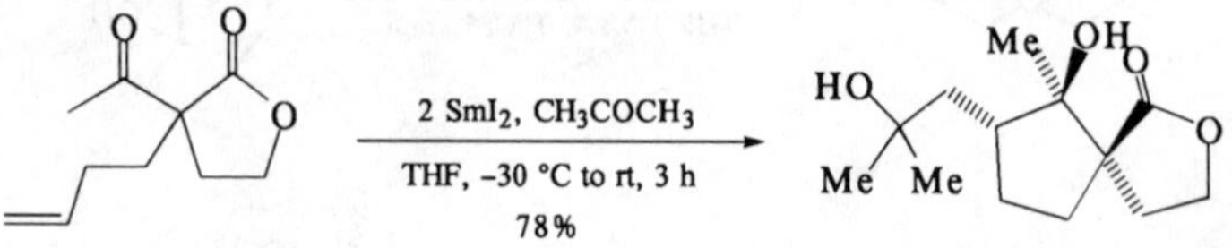

This protocol provides an efficient entry for the construction of cyclic alcohols. The revealed utility is seen in the synthesis of cyclopentanol with various functional groups. Furthermore, oxygen- or nitrogen-containing heterocyclic compounds can be synthesized by this cyclization reaction [4,12].

Elegant synthetic applications are illustrated below [13,14].

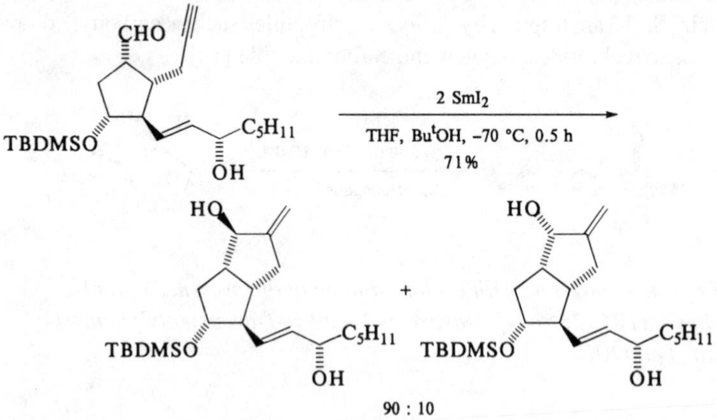

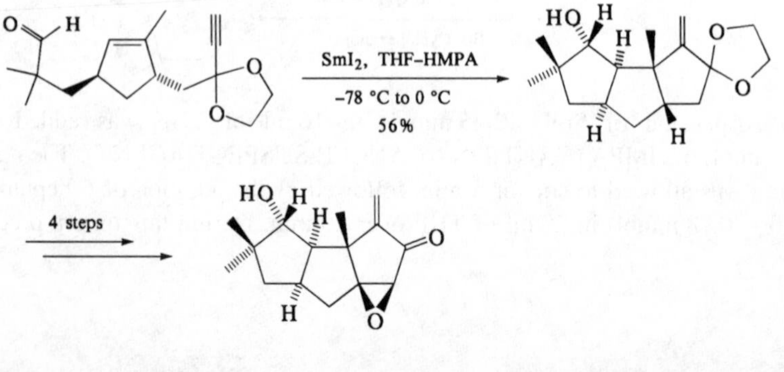

REFERENCES

1. S. Fukuzawa, M. Iida, A. Nakanishi, T. Fujinami and S. Sakai, *J. Chem. Soc. Chem. Commun.* 920 (1987).
2. S. Fukuzawa, A. Nakanishi, T. Fujinami and S. Sakai, *J. Chem. Soc. Perkin Trans. 1,* 1669 (1988).
3. E. J. Enholm and A. Trivellas, *Tetrahedron Lett.* **30,** 1063 (1989).
4. S. C. Shim, J.-T. Hwang, H.-Y. Kang and M. H. Chang, *Tetrahedron Lett.* **31,** 4765 (1990).
5. T. Gillmann, *Tetrahedron Lett.* **34,** 607 (1993).
6. E. J. Enholm and A. Trivellas, *J. Am. Chem. Soc.* **111,** 6463 (1989).
7. E. J. Enholm, H. Satici and A. Trivellas, *J. Org. Chem.* **54,** 5841 (1989).
8. M. Kito, T. Sakai, K. Yamada, F. Matsuda and H. Shirahama, *Synlett* 158 (1993).
9. G. A. Molander and C. Kenny, *Tetrahedron Lett.* **28,** 4367 (1987).
10. G. A. Molander and J. A. McKie, *J. Org. Chem.* **57,** 3132 (1992).
11. G. A. Molander and C. Kenny, *J. Org. Chem.* **56,** 1439 (1991).
12. J. E. Baldwin, S. C. M. Turner and M. G. Moloney, *Tetrahedron Lett.* **33,** 1517 (1992).
13. K. Bannai, T. Tanaka, N. Okamura, A. Hazato, S. Sugiura, K. Manabe, K. Tomimori, Y. Kato, S. Kurozumi and R. Noyori, *Tetrahedron* **46,** 6689 (1990).
14. T. L. Fevig, R. L. Elliott and D. P. Curran, *J. Am. Chem. Soc.* **110,** 5064 (1988).

4.4.3 Other radical cyclization reactions

One electron transfer from SmI_2 to organic halides generates radical species. These radicals undergo addition to unsaturated bonds to form carbon–carbon bonds. As shown in the following scheme, intramolecular addition to form five-membered rings occurs most readily, and the generated organosamarium(III) intermediates are trapped by a variety of electrophiles. By these cyclizations, oxygen and nitrogen heterocycles such as benzofurans and indole derivatives can be synthesized [1–6].

X – Br, I; Y – O, NR

Electrophile – R^1COR^2, H_2O, I_2, $(PhS)_2$, $(PhSe)_2$, Bu_3SnI, PhNCO

Preparation of 2H, 3H-3-(2-hydroxy-2-propylpentyl)benzofuran [5]

To an oven-dried flask equipped with a stir bar was added samarium metal powder (0.165 g, 1.1 mmol). The flask was flamed while being flushed with argon. The flask was cooled to 0 °C and 20 ml of THF added, followed by CH_2I_2 (0.267 g, 1.0 mmol). The solution remained at 0 °C for 15 min and was then warmed to room temperature and stirred for an additional hour. At this stage, 1.0 ml of HMPA (CAUTION – CANCER SUSPECT AGENT) was added, and the initially blue solution turned deep purple. A solution of 2-(allyloxy)-1-iodobenzene (0.104 g, 0.4 mmol) and 4-heptanone (0.055 g, 0.48 mmol) in 5 ml of THF was added to the SmI_2/HMPA solution over a period of approximately 1 min. After being stirred for 1 h, the reaction mixture was quenched with saturated K_2CO_3, followed by extraction with ether. The organic layer was washed twice with water, dried over $MgSO_4$, filtered and concentrated. The crude reaction product was purified by flash chromatography on silica gel (hexane–ethyl acetate, 7 : 1) to give 2H,3H-3-(2-hydroxy-2-propyl-pentyl)benzofuran (0.080 g, 81%) as a clear, colourless oil.

Other examples of radical cyclization reactions promoted by SmI_2 are shown below [7,8].

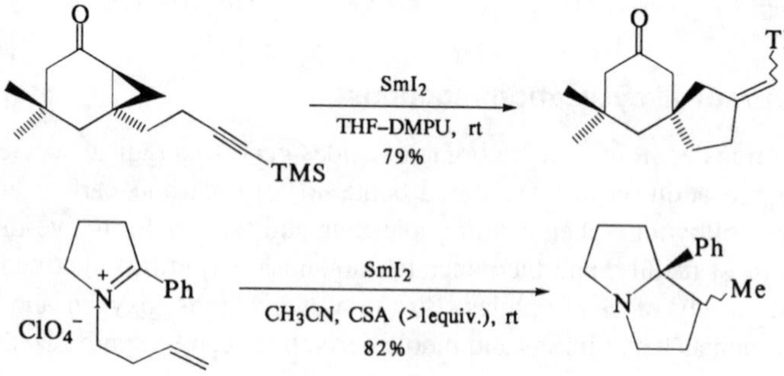

REFERENCES

1. D. P. Curran, T. L. Fevig, C. P. Jasperse and M. J. Totleben, *Synlett* 943 (1992).
2. D. P. Curran, T. L. Fevig and M. J. Totleben, *Synlett* 770 (1990).
3. D. P. Curran and M. J. Totleben, *J. Am. Chem. Soc.* **114**, 6050 (1992).
4. M. J. Totleben, D. P. Curran and P. Wipf, *J. Org. Chem.* **57**, 1740 (1992).
5. G. A. Molander and L. S. Harring, *J. Org. Chem.* **55**, 6171 (1990).
6. J. Inanaga, O. Ujikawa and M. Yamaguchi, *Tetrahedron Lett.* **32**, 1737 (1991).
7. R. A. Batey and W. B. Motherwell, *Tetrahedron Lett.* **32**, 6649 (1991).
8. S. F. Martin, C.-P. Yang, W. L. Laswell and H. Rüeger, *Tetrahedron Lett.* **29**, 6685 (1988).

4.4.4 Miscellaneous reactions

Aryl radicals generated by the reaction of aryl halides with SmI_2 can be utilized for new and efficient carbon–carbon bond formation. Masked formylation of carbonyl compounds in the presence of iodobenzene and 1,3-dioxolane is a typical example [1]. This reaction involves a 1,3-dioxolanyl radical intermediate which is generated by the reaction of 1,3-dioxolane with the initially formed phenyl radical.

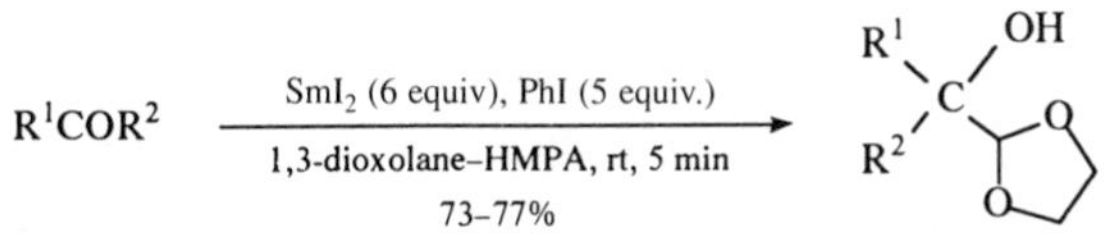

$$R^1COR^2 \quad \xrightarrow[\substack{1,3\text{-dioxolane–HMPA, rt, 5 min} \\ 73\text{–}77\%}]{\text{SmI}_2 \text{ (6 equiv), PhI (5 equiv.)}} \quad$$

Another example involving an aryl radical intermediate is shown below [2].

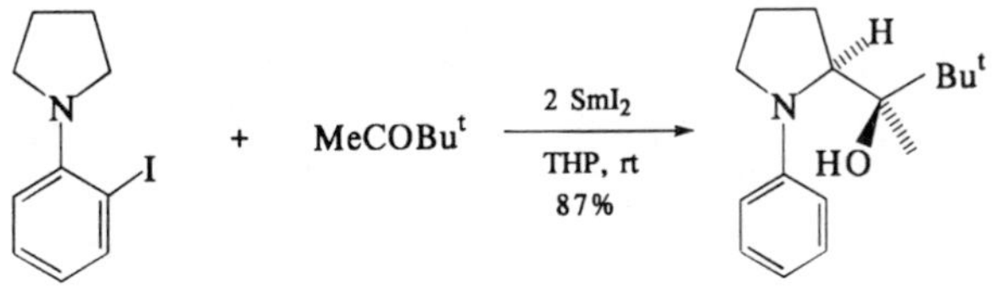

α,β-Unsaturated esters are instantaneously dimerized at room temperature in either intermolecular or intramolecular fashion by the use of reduction system SmI$_2$–THF–HMPA in the presence of a proton source [3]. Conjugated dienoic esters are also dimerized at the γ-position with the migration of the remaining double bonds.

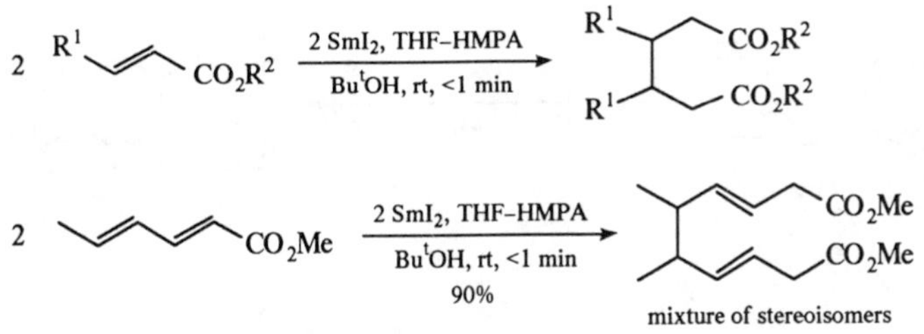

mixture of stereoisomers

Under similar conditions, the reaction of *N,N*-dibenzyl crotonamide affords the corresponding (±)-3,4-dimethyladipamide as a single stereoisomer [3]

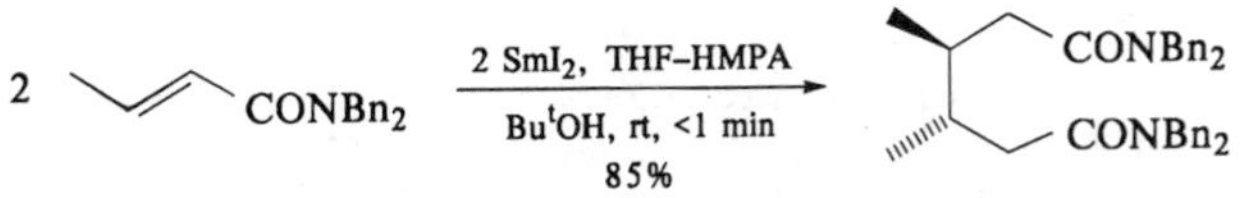

Samarium carbenoids, generated by the reaction of organic halides with SmI$_2$, add to alkenes, as exemplified by the following schemes [4,5].

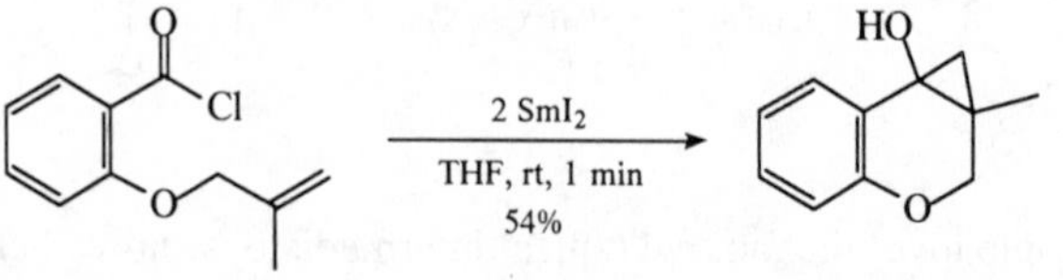

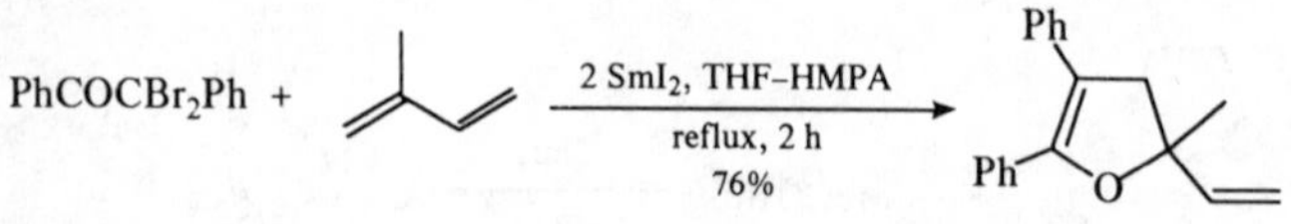

2,3-Naphthoquinodimethane is generated from o-bis(α-acetoxypropargyl)-benzene by treatment with SmI_2 in the presence of Pd(0) catalyst [6]. The quinodimethane intermediate reacts with various dienophiles to afford the corresponding Diels–Alder adducts in good yields.

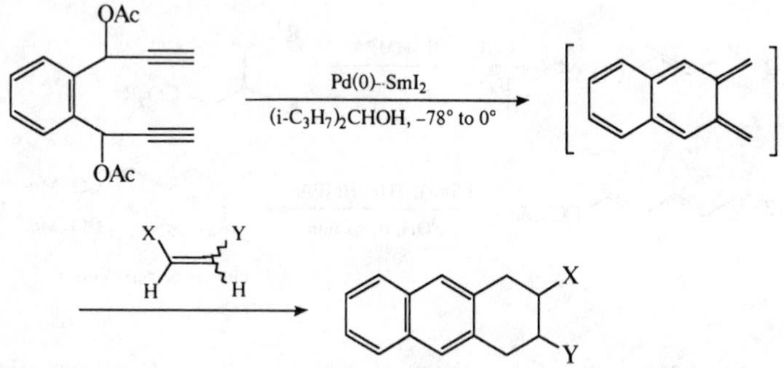

REFERENCES

1. M. Matsukawa, J. Inanaga and M. Yamaguchi, *Tetrahedron Lett.* **28,** 5877 (1987).
2. M. Murakami, M. Hayashi and Y. Ito, *J. Org. Chem.* **57,** 793 (1992).
3. J. Inanaga, Y. Handa, T. Tabuchi, K. Otsubo, M. Yamaguchi and T. Hanamoto, *Tetrahedron Lett.* **32,** 6557 (1991).
4. M. Sasaki, J. Collin and H. B. Kagan, *Tetrahedron Lett.* **29,** 6105 (1988).
5. S. Fukuzawa, T. Fujinami and S. Sakai, *J. Chem. Soc. Chem. Commun.* 919 (1987).
6. J. Inanaga, Y. Sugimoto and T. Hanamoto, *Tetrahedron Lett.* **33,** 7035 (1992).

4.5 ORGANIC REACTIONS WITH DICYCLOPENTADIENYLSAMARIUM(II) AND BIS(PENTAMETHYLCYCLOPENTADIENYL)SAMARIUM(II)

The number of organic reactions with dicyclopentadienylsamarium(II), $(C_5H_5)_2Sm$, and bis(pentamethylcyclopentadienyl)samarium(II), $(C_5Me_5)_2Sm$, are at present limited. However, these complexes have different properties from SmI_2, and their reported reactions are unique, some of them being conceptually new. Therefore, these divalent samarium complexes are potentially useful as reagents and catalysts in organic synthesis.

4.5.1 Dicyclopentadienylsamarium(II)

Dicyclopentadienylsamarium(II) promotes the Barbier-type reaction between alkyl halides and carbonyl compounds [1]. The reaction provides cross-coupling products in higher yields than the reaction using SmI_2, probably because $(C_5H_5)_2Sm$ suppresses undesirable side reactions such as the Meerwein–Ponndorf–Verley–Oppenauer reaction and the Tishchenko reaction. A distinct difference in the reactivities toward organic halides is the $(C_5H_5)_2Sm$ reacts with benzylic or allylic halides to provide benzylic and allylic samarium(III) complexes which are not accessible by the use of SmI_2 [2–4]. The generated complexes react with carbonyl compounds and acid chlorides to give alcohols and ketones, respectively.

$$ArCH_2Cl \quad + \quad 2\ Sm(C_5H_5)_2 \quad \xrightarrow[\text{rt}]{\text{THF}} \quad ArCH_2Sm(C_5H_5)_2 \quad + \quad ClSm(C_5H_5)_2$$

$$ArCH_2Sm(C_5H_5)_2 \quad \xrightarrow{R^1COR^2} \quad ArCH_2-\underset{\underset{R^1}{|}}{\overset{\overset{OH}{|}}{C}}-R^2$$

Ar = Ph, 4-Bu^{t}C$_6$H$_4$, 2,3-(CH$_3$)$_2$C$_6$H$_3$

An interesting carbonylation reaction occurs on treatment of t-butyl bromide with $(C_5H_5)_2Sm$ under carbon monoxide [5]. An acylsamarium complex may be a reactive intermediate in this reaction.

$$Bu^tBr \quad \xrightarrow[\text{THF, } -20\ ^\circ C \text{ to } +20\ ^\circ C]{CO,\ (C_5H_5)_2Sm\ (2.5\text{–}3.5\ \text{equiv.})}$$

$$Bu^t-\underset{\underset{OH}{|}}{CH}-\underset{\underset{OH}{|}}{CH}-Bu^t \quad + \quad Bu^t-\underset{\overset{||}{O}}{C}-\underset{\overset{||}{O}}{C}-\underset{\underset{OH}{|}}{CH}-Bu^t$$

REFERENCES

1. J. L. Namy, J. Collin, J. Zhang and H. B. Kagan, *J. Organometal. Chem.* **328,** 81 (1987).
2. J. Collin, J. L. Namy, C. Bied and H. B. Kagan, *Inorg. Chim. Acta* **140,** 29 (1987).

3.   J. Collin, C. Bied and H. B. Kagan, *Tetrahedron Lett.* **32**, 629 (1991).
4.   C. Bied, J. Collin and H. B. Kagan, *Tetrahedron* **48**, 3877 (1992).
5.   J. Collin and H. B. Kagan, *Tetrahedron Lett.* **29**, 6097 (1988).

4.5.2 Bis(pentamethylcyclopentadienyl)samarium(II)

The reactivities of $(C_5Me_5)_2Sm$ toward organic compounds have been extensively investigated by Evans and his coworkers. This complex, which is soluble in most organic solvents, reacts with various organic functional groups to give novel organometallic complexes [1–9]. Some of them are illustrated below. These reactions are extremely interesting not only as organometallic reactions but also from a synthetic point of view.

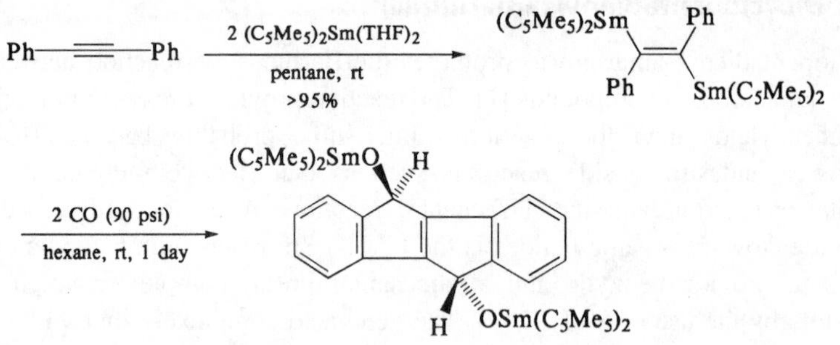

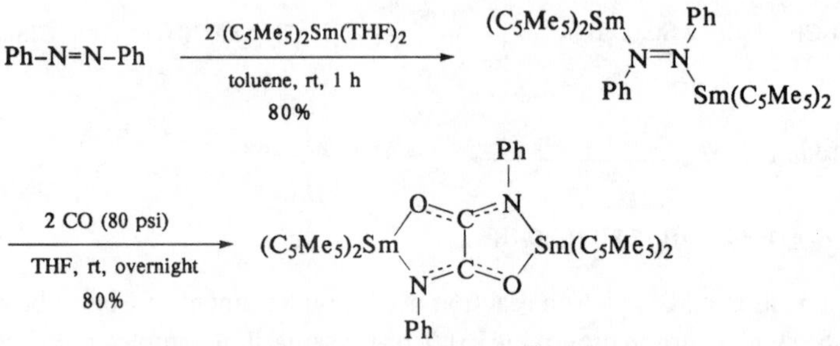

Reaction of 1,3-butadiene with bis(pentamethylcyclopentadienyl)samarium(II) [8]

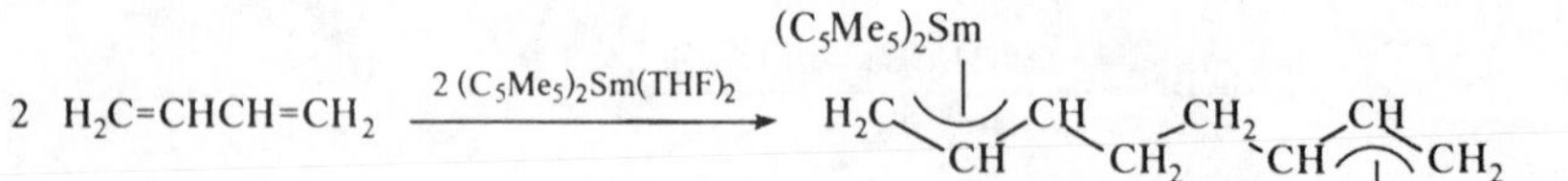

In a glove box, $(C_5Me_5)_2Sm(THF)_2$ (0.214 g, 0.38 mmol) was suspended in 10 ml of hexane to give a dark-purple suspension which was placed in a 100 ml round-bottom flask equipped with a gas inlet and a Teflon stir bar. The pressure in the reaction vessel was reduced until the solution bubbled. Butadiene (1 atm) was placed over the stirred solution causing an immediate colour change to deep red and causing all compounds to go into solution. During the next 3 h, a red-orange precipitate was deposited. Excess butadiene was removed, and the mixture was filtered to yield complex **8** as a red-orange microcrystalline powder (0.131 g, 73%).

Reaction of 1,2-di-2-pyridylethene with carbon monoxide in the presence of bis(pentamethylcyclopentadienyl)samarium(II) [9]

(9)

In a glove box, 1,2-di-2-pyridylethene (0.040 g, 0.22 mmol) in 4 ml of toluene was slowly added to $(C_5Me_5)_2Sm(THF)_2$ (0.250 g, 0.442 mmol) in 10 ml of toluene to form a red solution. The solution was placed in a 3-oz Fisher–Porter aerosol reaction vessel and pressurized with CO to 80 psi. After 24 h, the resulting yellow-orange solution was depressurized, concentrated to about half-volume, and cooled to −34 °C. Yellow-orange crystals of complex **8** (0.180 g, 65%) were obtained. The ^{1}H-NMR spectrum of the bulk reaction mixture indicated that compound **9** was the major product (90% yield)

REFERENCES

1. W. J. Evans, *Polyhedron* **6**, 803 (1987).
2. W. J. Evans, I. Bloom, W. E. Hunter and J. L. Atwood, *J. Am. Chem. Soc.* **105**, 1401 (1983).
3. W. J. Evans, L. A. Hughes, D. K. Drummond, H. Zhang and J. L. Atwood, *J. Am. Chem. Soc.* **108**, 1722 (1986).
4. W. J. Evans, D. K. Drummond, L. R. Chamberlain, R. L. Doedens, S. G. Bott, H. Zhang and J. L. Atwood, *J. Am. Chem. Soc.* **110**, 4983 (1988).
5. W. J. Evans, J. W. Grate, L. A. Hughes, H. Zhang and J. L. Atwood, *J. Am. Chem. Soc.* **107**, 3728 (1985).
6. W. J. Evans, T. A. Ulibarri and J. W. Ziller, *J. Am. Chem. Soc.* **112**, 219 (1990).
7. W. J. Evans and D. K. Drummond, *J. Am. Chem. Soc.* **111**, 3329 (1989).
8. W. J. Evans, T. A. Ulibarri and J. W. Ziller, *J. Am. Chem. Soc.* **112**, 2314 (1990).
9. W. J. Evans and D. K. Drummond, *J. Am. Chem. Soc.* **110**, 2772 (1988).

–5–

Trivalent Lanthanides

The trivalent state is the most common oxidation state for lanthanides. Lanthanides (III) are considered to be hard acids according to Pearson's HSAB classification, being located between Mg(II) and Ti(IV). Therefore, lanthanide(III) species have strong affinity toward hard bases such as oxygen donor ligands. This strong oxophilicity is one of the most important characteristic features of lanthanide(III) compounds and is often utilized for selective functional group transformations.

5.1 CARBONYL REDUCTIONS PROMOTED BY LANTHANIDE(III) REAGENTS

5.1.1 Reduction of ketones by NaBH$_4$–CeCl$_3$

5.1.1.1 Selective 1,2-reduction of α-enones

Reduction of carbonyl compounds by NaBH$_4$ is a versatile synthetic reaction, but reduction with NaBH$_4$ alone is often accompanied by undesirable side reactions. For example, reduction of α-enones by NaBH$_4$ alone is usually accompanied by 1,4-reduction, affording a mixture of allylic alcohols and saturated alcohols. In 1978, Luche reported the effectiveness of lanthanide salts in selective reduction of carbonyl compounds by NaBH$_4$. The most conspicuous advantage is seen in the selective 1,2-reduction of α-enones [1–3]. Cerium chloride exhibits the highest efficacy, although other lanthanide salts are also highly efficient.

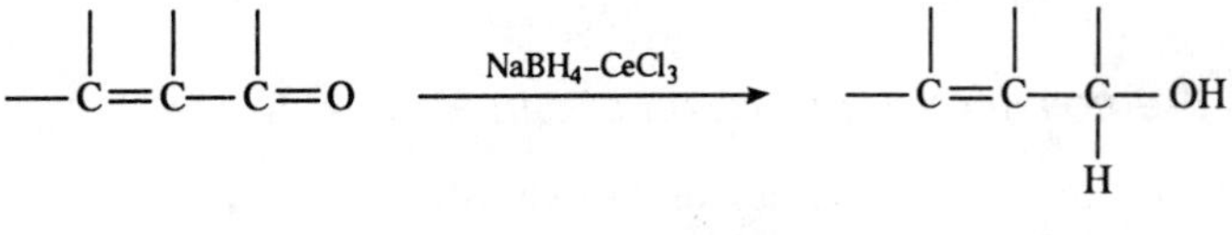

$$-\underset{|}{C}=\underset{|}{C}-\underset{|}{C}=O \quad \xrightarrow{\text{NaBH}_4\text{–CeCl}_3} \quad -\underset{|}{C}=\underset{|}{C}-\underset{\underset{H}{|}}{C}-OH$$

The reaction mechanism accounting for the regioselectivity was proposed by the authors on the basis of the Pearson's hard–soft principle [3]. Sodium borohydride is rapidly converted to the $[BH_{4-n}^-(OMe)_n]$ species in methanol by the action of $CeCl_3$. On the other hand, the carbonyl function is activated by hydrogen bonding as depicted below, the carbonyl carbon being harder than the non-activated one. The hard borohydride prefers to attack the hard carbonyl carbon to give allylic alcohols.

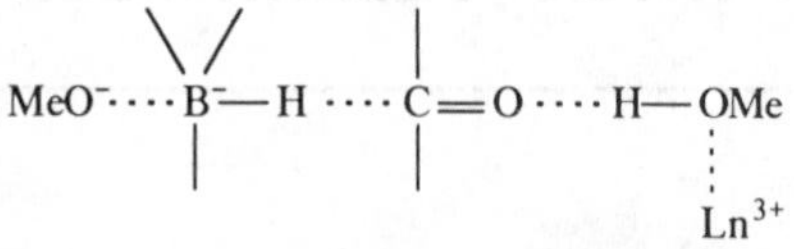

$$MeO^- \cdots B^-\!\!-\!\!H \cdots C\!\!=\!\!O \cdots H\!\!-\!\!OMe$$
$$Ln^{3+}$$

The utility of this method has been proved by numerous applications in practical organic synthesis. Table 5.1 shows representative examples. Not only regioselectivity but also chemo- and stereoselectivity are illustrated in these examples.

Reduction of 2-cyclopentenone by NaBH$_4$–CeCl$_3$ [14]

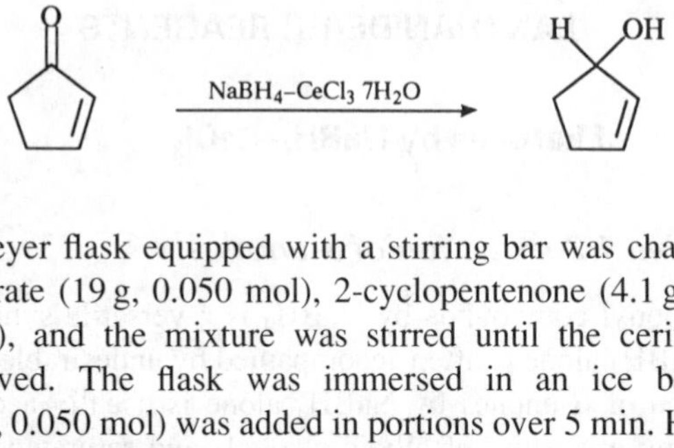

A 300 ml Erlenmeyer flask equipped with a stirring bar was charged with cerium chloride heptahydrate (19 g, 0.050 mol), 2-cyclopentenone (4.1 g, 0.050 mol) and methanol (120 ml), and the mixture was stirred until the cerium chloride was completely dissolved. The flask was immersed in an ice bath, and sodium borohydride (1.9 g, 0.050 mol) was added in portions over 5 min. Hydrogen evolved vigorously with elevation of temperature to 15–20°C. After the mixture was stirred for an additional 10 min, the resulting white suspension was concentrated to about 30 ml under reduced pressure. Water (100 ml) was added to dissolve inorganic salts and the mixture was extracted with ether (3 × 50 ml). The combined extracts were dried over $MgSO_4$ and concentrated to a volume of *ca.* 10 ml. Distillation under ordinary pressure afforded 2-cyclopenten-1-ol (3.7 g, 88%) with 96% purity.

5.1.1.2   Stereoselective reduction of saturated ketones

The oxophilicity of Ce^{3+} allows stereoselective reduction of saturated ketones. Table 5.2 shows some representative examples. The stereochemical outcomes can be reasonably explained in terms of chelation control. Cheap reagents and the simplicity of the experimental procedure enhance the usefulness of this method.

TABLE 5.1

Reductions of α-Enones by NaBH$_4$–CeCl$_3$

Substrate	Conditions	Product(s) (yield)	%Ref.
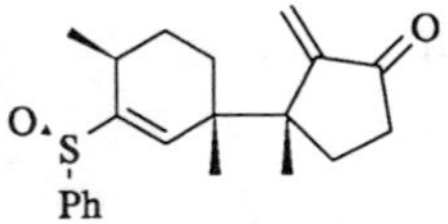 (structure: $O\!-\!S\!-\!Ph$ cyclohexenone–methylenecyclopentanone)	MeOH, –10 °C	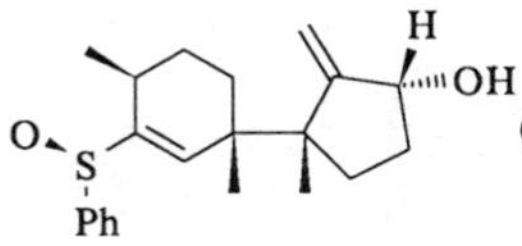 (structure, OH/H, $\overset{\cdot}{O}\!-\!S\!-\!Ph$) (83) 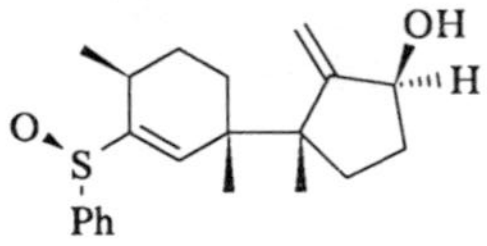 and (structure, OH/H) (9)	[4]
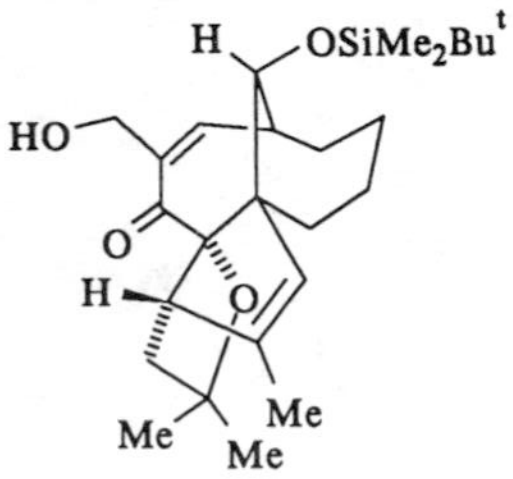 (structure: $OSiMe_2Bu^t$, HO)	MeOH, 0 °C	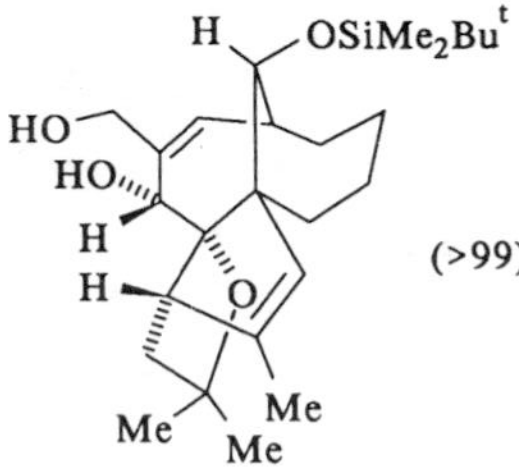 (structure: H $OSiMe_2Bu^t$, HO, Me Me Me) (>99)	[5]
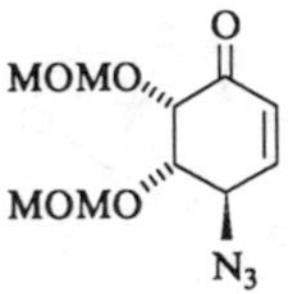 (structure: MOMO, MOMO, N_3)	MeOH, rt	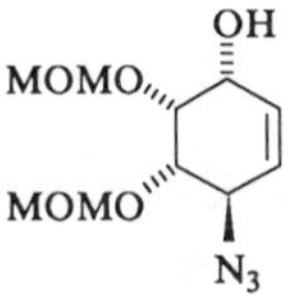 (structure: OH, MOMO, MOMO, N_3) (86)	[6]
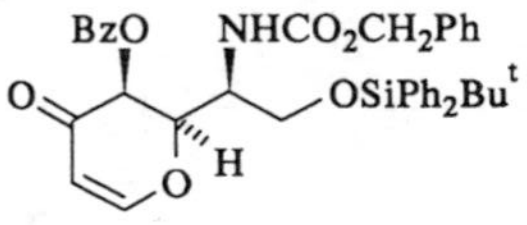 (structure: BzO, $NHCO_2CH_2Ph$, $OSiPh_2Bu^t$, H, O)	MeOH, –78 °C to rt	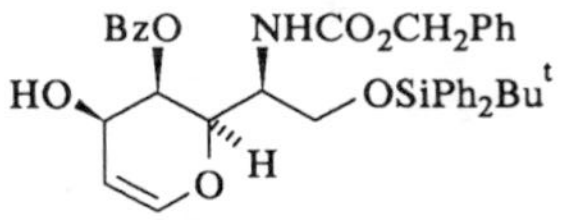 (structure: $NHCO_2CH_2Ph$, $OSiPh_2Bu^t$, BzO, HO, O) (97)	[7]

continued

TABLE 5.1 *continued*

Reductions of α-Enones by NaBH$_4$–CeCl$_3$

Substrate	Conditions	Product(s) (yield %)	Ref.
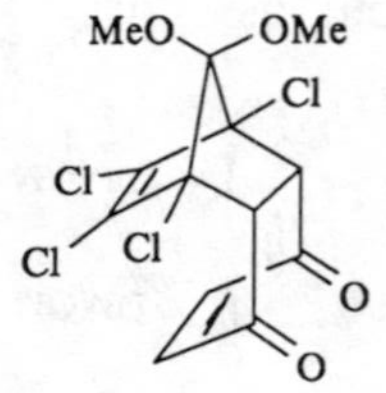 MeO OMe, Cl, Cl, Cl (structure)	MeOH, rt	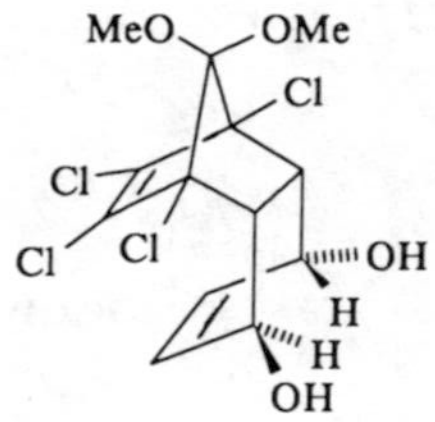 OH, Cl, MeO OMe, Cl, Cl, Cl, H, OH (60)	[8]
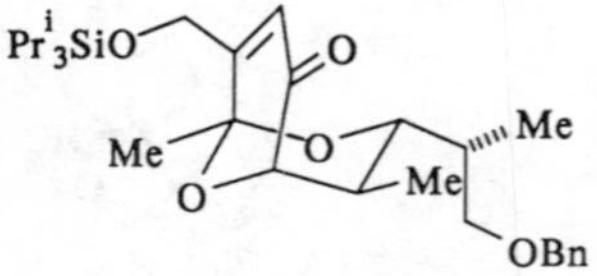 Pr$^i_3$SiO, Me, Me, OBn, Me (structure)	MeOH, rt	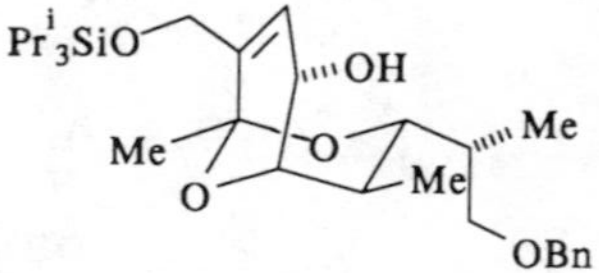 Pr$^i_3$SiO, OH, Me, Me, OBn, Me (99)	[9]
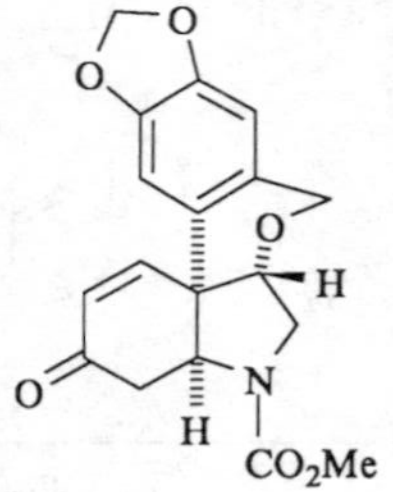 (methylenedioxy, O, H, N—CO$_2$Me, H, O) (structure)	MeOH, −78 °C	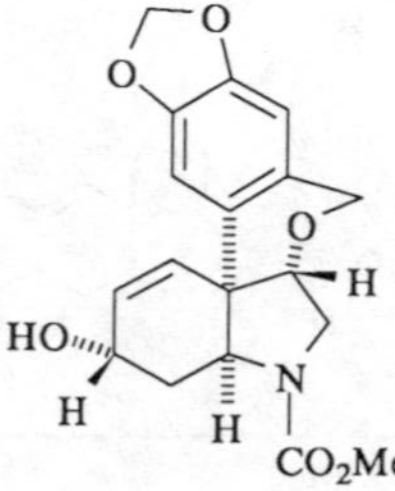 (methylenedioxy, O, H, N—CO$_2$Me, HO, H, H) (92) 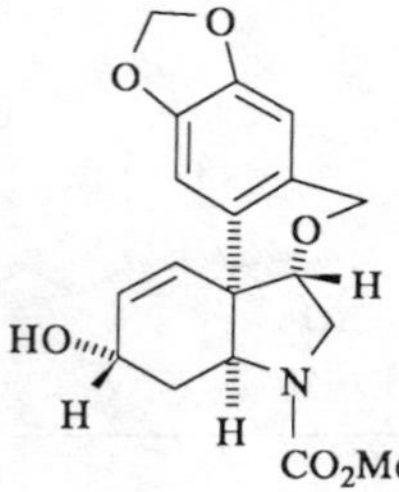 and (3)	[10]

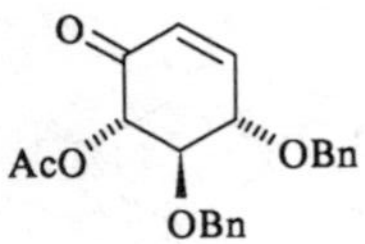O AcO OBn OBn	MeOH, 0 °C	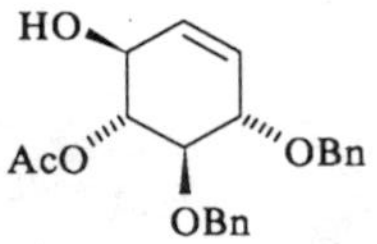HO AcO OBn OBn (86)	[11]
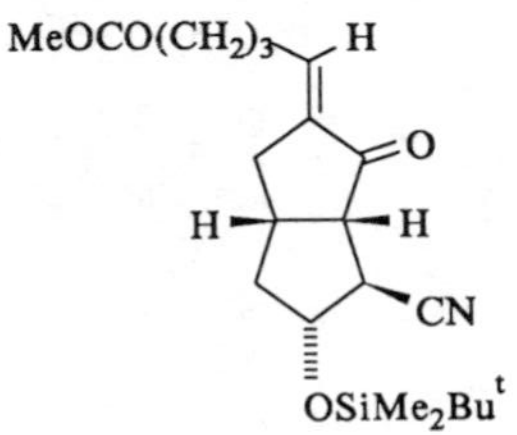MeOCO(CH2)3 H O H H CN OSiMe2But	MeOH, rt	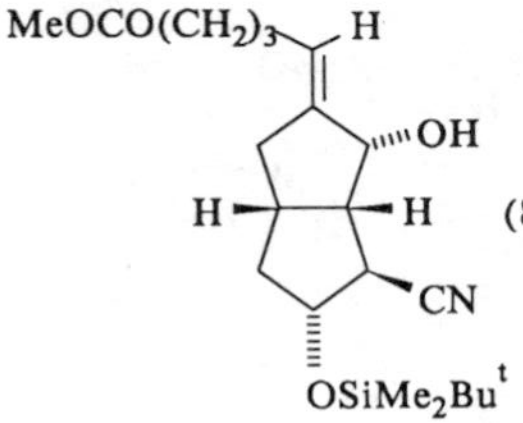MeOCO(CH2)3 H OH H H CN OSiMe2But (80) 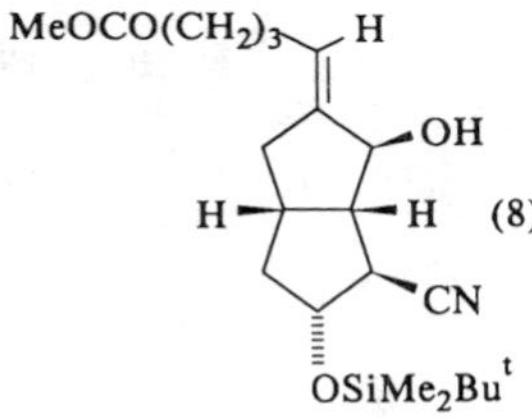MeOCO(CH2)3 H OH H H CN OSiMe2But (8)	[12]
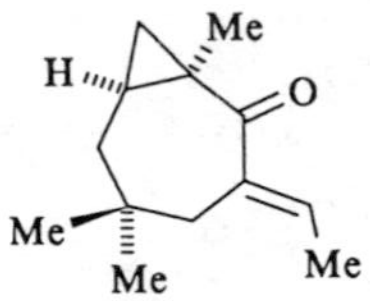H Me O Me Me Me	MeOH, rt	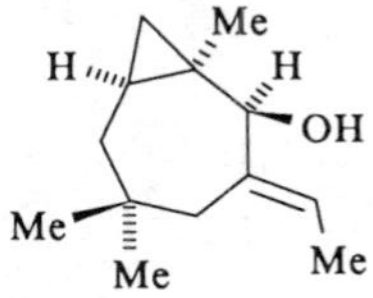H Me H OH Me Me Me (97)	[13]

TABLE 5.2

Stereoselective Reductions of Ketones by NaBH$_4$–CeCl$_3$

Substrate	Conditions	Product(s) (yield %)	Ref.
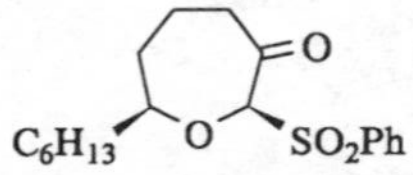 (structure with SO$_2$Ph, C$_6$H$_{13}$)	MeOH, rt	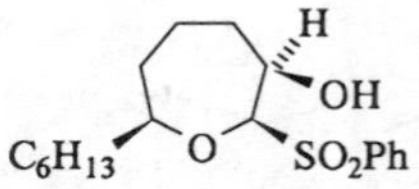 (97)	[15]
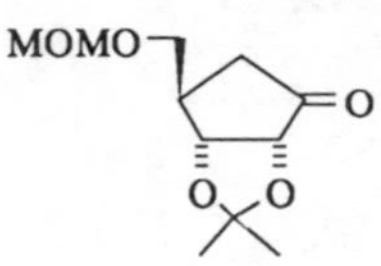 (MOMO structure)	MeOH, rt	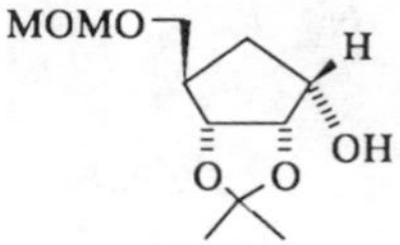 (92)	[16]
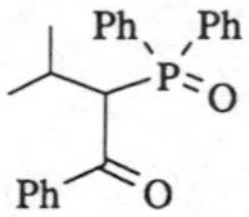 (Ph$_2$P=O structure)	EtOH, −78 °C	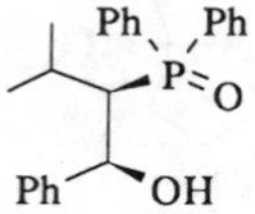 (>95) 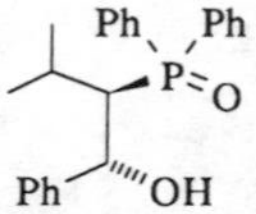 (<5)	[17]
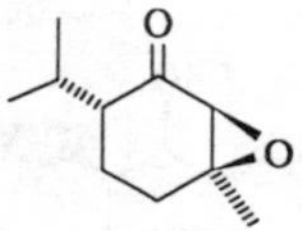 (epoxide structure, isopropyl)	MeOH, rt	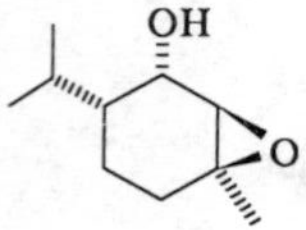 (92) 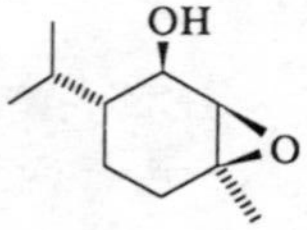 (3)	[18]

5.1.1.3   Selective reduction of ketones in the presence of aldehydes

The cerium ion is an effective catalyst for the acetalization or hemiacetalization of aldehydes in an alcoholic medium (see section 5.5.1), and hence aldehyde is protected from nucleophilic attack under these conditions. In contrast, the ketone group is not subjected to acetalization. Therefore, selective reduction of ketones in the presence of aldehydes by $NaBH_4$ is possible in the presence of cerium chloride in alcoholic or aqueous medium [19,20].

This method is applicable to the selective reduction of saturated alkyl aldehydes in the presence of α,β-unsaturated aldehydes or aromatic aldehydes [21]. It should be noted that the less reactive carbonyl groups are preferably reduced leaving the more reactive ones intact. This method of achieving reverse selectivity is particularly attractive, since it is carried out in one flask without the need for any prior protection.

REFERENCES

1.   J.-L. Luche, *J. Am. Chem. Soc.* **100**, 2226 (1978).
2.   J.-L. Luche, L. Rodriguez-Harn and P. Crabbe, *J. Chem. Soc. Chem. Commun.* 601 (1978).
3.   A. L. Gemal and J.-L. Luche, *J. Am. Chem. Soc.* **103**, 5454 (1981).
4.   D. H. Hua, S. Venkataraman, R. Chan-Yu-King and J. V. Paukstelis, *J. Am. Chem. Soc.* **110**, 4741 (1988).
5.   L. A. Paquette, R. J. Ross and J. P. Springer, *J. Am. Chem. Soc.* **110**, 6192 (1988).
6.   N. Chida, M. Ohtsuka and S. Ogawa, *Tetrahedron Lett.* **32**, 4525 (1991).
7.   A. Golebiowski, J. Kozak and J. Jurczak, *J. Org. Chem.* **56**, 7344 (1991).
8.   A. P. Marchand, W. D. LaRoe, G. V. M. Sharma, S. C. Suri and D. S. Reddy, *J. Org. Chem.* **51**, 1622 (1986).
9.   S. J. Shimshock, R. E. Waltermire and P. DeShong, *J. Am. Chem. Soc.* **113**, 8791 (1991).
10.   M. M. Abelman, L. E. Overman and V. D. Tran, *J. Am. Chem. Soc.* **112**, 6959 (1990).
11.   N. Chida, M. Ohtsuka, K. Nakazawa and S. Ogawa, *J. Org. Chem.* **56**, 2976 (1991).
12.   I. Fleming and D. Higgins, *Tetrahedron Lett.* **30**, 5777 (1989).
13.   L. A. Paquette and W. H. Ham, *J. Am. Chem. Soc.* **109**, 3025 (1987).
14.   T. Imamoto and T. Hatajima, in *Practical Organometallic Chemistry for Synthetic Chemists*, ed. F. Sato, K. Yamamoto and T. Imamoto, p. 223. Kodansha, Tokyo, 1992 (reproduced by permission from T. Imamoto and Kodansha).

15. M. J. Davies, C. J. Moody and R. J. Taylor, *J. Chem. Soc. Perkin Trans. 1* (1991).
16. H. J. Bestmann and D. Roth, *Synlett* 751 (1990).
17. J. Elliott, D. Hall and S. Warren, *Tetrahedron Lett.* **30**, 601 (1989).
18. G. Rücker, H. Hörster and W. Gajewski, *Synth. Commun.* **10**, 623 (1980).
19. J.-L. Luche and A. L. Gemal, *J. Am. Chem. Soc.* **101**, 5848 (1979).
20. A. L. Gemal and J-L. Luche, *J. Org. Chem.* **44**, 4187 (1979).
21. A. L. Gemal and J.-L. Luche, *Tetrahedron Lett.* **22**, 4077 (1981).

5.1.2   Meerwein–Ponndorf–Verley reductions and Tishchenko reactions

Lanthanide   alkoxides   exhibit   efficient   catalytic   activity   in   the Meerwein–Ponndorf–Verley reduction [1–4]. Thus, various aldehydes and ketones can be reduced to alcohols by treatment with an excess amount of i-propanol in the presence of $Ln(OPr^i)_3$ or $Ln(OR)I_2$. These catalysts are far more reactive than other existing catalysts. For example, the activity of $Gd(OPr^i)_3$ is about 10^3 times as high as that of $Al(OPr^i)_3$ [4]. These catalysts, however, are extremely moisture sensitive as well as the other alkoxides and therefore should be handled under anhydrous conditions. The catalysts, $Sm(OR)I_2$, are prepared *in situ* by SmI_2-promoted Barbier-type reaction of alkyl iodides with carbonyl compounds or by the reaction of SmI_2 with di-t-butylperoxide [1]. Representative examples of lanthanide alkoxide-catalysed Meerwein–Ponndorf–Verley reductions are shown in Table 5.3.

The utility of these methods has been demonstrated by the ketone reductions in the synthesis of rutamycin spiroketal synthon and cytovaricin [5,6]. It should be noted that other methods such as use of $LiAlH_4$ or employment of dissolving metal conditions result in the formation of the undesired stereoisomeric alcohol or 1:1 mixtures of the stereoisomers.

A very elegant Tishchenko reduction has been achieved by the use of samarium alkoxide as a catalyst [7]. Various β-hydroxy ketones, on treatment with excess aldehydes in the presence of SmI_2 (15 mol %), are reduced to *anti* 1,3-diol monoesters in excellent yields with almost complete stereoselectivity. The reaction involves initial hemiacetal formation and subsequent intramolecular hydride transfer. The excellent level of stereochemical control is attributed to the coordination of samarium to both carbonyl and hemiacetal oxygen.

TABLE 5.3

Meerwein–Ponndorf–Verley Reduction Catalysed by Lanthanide Alkoxides[a]

Substrate	Catalyst (mol %)	Conditions	Product(s) (yield %)	Ref.
2-Octanone	La(OPr^i)$_3$ (5)	80 °C, 18h	2-Octanol (>98)	[3]
2-Octanone	Ce(OPr^i)$_3$ (10)	80 °C, 24h	2-Octanol (95)	[3]
2-Octanone	Sm(OPr^i)$_3$ (10)	80 °C, 18h	2-Octanol (>98)	[3]
2-Octanone	Gd(OPr^i)$_3$ (2)	80 °C, 2h	2-Octanol (85)	[4]
2-Octanone	Yb(OPr^i)$_3$ (5)	80 °C, 24h	2-Octanol (40)	[3]
2-Octanone	Bu^{t}OSmI$_2$ (10)	65 °C, 24h	2-Octanol (86)	[1]
Cyclohexanone	Nd(OPr^i)$_3$ (1)	30 °C, 0.5h	Cyclohexanol (73)	[4]
Cyclohexanone	Gd(OPr^i)$_3$ (1)	30 °C, 0.5h	Cyclohexanol (99)	[4]
(E)-4-phenyl-3-buten-2-one	Gd(OPr^i)$_3$ (2)	80 °C, 2h	(E)-4-phenyl-3-buten-2-ol (73)	[4]
Acetophenone	Gd(OPr^i)$_3$ (2)	80 °C, 2h	α-Methylbenzyl alcohol (82)	[4]
3-Nitroacetophenone	Gd(OPr^i)$_3$ (2)	80 °C, 2h	3-Nitro-α-methylbenzyl alcohol (97)	[4]
Benzophenone	Nd(OPr^i)$_3$ (2)	80 °C, 2h	Benzhydrol (85)	[4]
Octanal	Bu^{t}O$_S$mI$_2$ (10)	65 °C, 24h	1-Octanol (66) Octanoic acid (33)	[1]
i-Butanal	Gd(OPr^i)$_3$ (2)	30 °C, 0.1h	i-Butanol (74)	[4]
Benzaldehyde	Gd(OPr^i)$_3$ (2)	0 °C, 2h	Benzyl alcohol (52)	[4]
4-Nitrobenzaldehyde	Bu^{t}OSmI$_2$ (10)	65 °C, 24h	4-Nitrobenzyl alcohol (94)	[1]
4-Methoxybenzaldehyde	Bu^{t}OSmI$_2$ (10)	65 °C, 24h	4-Methoxybenzyl alcohol (58) 4-Methoxybenzoic acid (29)	[1]
Ethyl pyruvate	Bu^{t}OSmI$_2$ (10)	65 °C, 24h	Ethyl lactate (80)	[1]

[a]All reactions are carried out using i-propanol (4–20 equivalents) as a reducing agent.

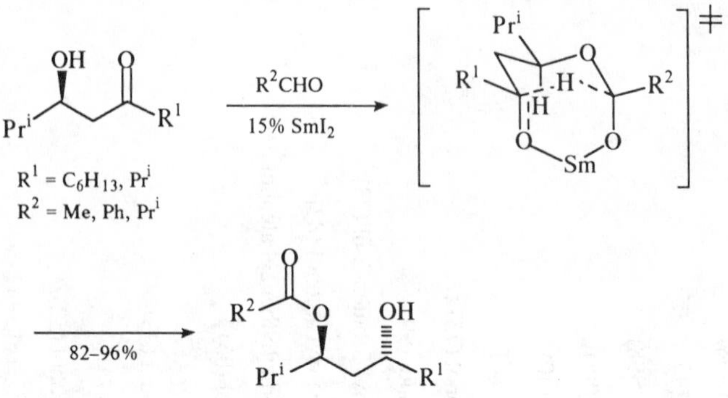

$R^1 = C_6H_{13}, Pr^i$
$R^2 = Me, Ph, Pr^i$

anti : syn = > 99 : 1

Synthesis of (3R,5R*)-5-benzoyloxy-2,6-dimethyl-3-heptanol [7]*

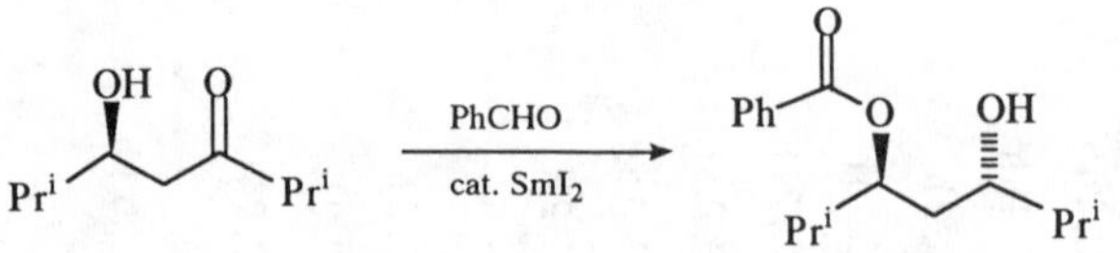

2,6-Dimethyl-5-hydroxy-3-heptanone (2.58 g, 16.3 mmol) was dissolved in anhydrous THF (45 ml) under nitrogen. To this solution was added 6.90 ml (65.3 mmol) of freshly distilled benzaldehyde. The solution was cooled to −10 °C (methanol–ice bath) and covered with aluminium foil, and 24.5 ml (2.45 mmol) of a 0.1 M stock solution of SmI_2 in THF was added dropwise (1 mm). The initial blue colour disappeared within 15 s. After 10 min, the reaction mixture was quenched by the addition of diethyl ether and saturated aqueous $NaHCO_3$. The ethereal solution was washed with saturated aqueous $NaHCO_3$, dried and concentrated. Silica gel chromatography (hexane–ether, 6 : 1) afforded (3R*,5R*)-5-benzoyloxy-2,6-dimethyl-3-heptanol (4.28 g, 99%) as a pale-yellow oil with an isomeric purity of more than 99%.

The intramolecular Tishchenko reaction is also achieved by the use of samarium alkoxide. A representative reaction providing a single stereoisomer is shown below [8].

OBn
Me
Me   CHO   cat. SmI$_2$(OBu^t)
TBDMSO''''   O   THF, rt
           89%
Me

OBn
Me   O
Me
TBDMSO''''   O
Me

REFERENCES

1. J. L. Namy, J. Souppe, J. Collin and H. B. Kagan, *J. Org. Chem.* **49**, 2045 (1984).
2. J. Collin, J. L. Namy and H. B. Kagan, *Nouv. J. Chim.* **10**, 229 (1986).
3. A. Lebrun, J. L. Namy and H. B. Kagan, *Tetrahedron Lett.* **32**, 2355 (1991).
4. T. Okano, M. Matsuoka, H. Konishi and J. Kiji, *Chem. Lett.* **181**, (1987).
5. D. A. Evans, D. L. Rieger, T. K. Jones and S. W. Kaldor, *J. Org. Chem.* **55**, 6260 (1990).
6. D. A. Evans, S. W. Kaldor, T. K. Jones, J. Clardy and T. J. Stout, *J. Am. Chem. Soc.* **112**, 7001 (1990).
7. D. A. Evans and A. H. Hoveyda, *J. Am. Chem. Soc.* **112**, 6447 (1990).
8. J. Uenishi, S. Masuda and S. Wakabayashi, *Tetrahedron Lett.* **32**, 5097 (1991).

5.2 OLEFIN HYDROGENATION CATALYSED BY LANTHANIDE (III) COMPLEXES

Organolanthanide(III) complexes having pentamethylcyclopentadienyl and related moieties catalyse olefin hydrogenation [1,2]. Hydrogenation with these catalysts proceeds under homogeneous conditions, and the activities of these catalysts for hydrogenation of simple terminal olefins are exceedingly high compared with other homogeneous catalysts of d-block transition elements. For example, a turnover N_t of 120 000 h^{-1} for $((C_5Me_5)_2LuH)_2$-catalysed hydrogenation of 1-hexene at 25 °C, 1 atm of H$_2$ pressure, is compared with reported N_t values of 650 h^{-1} for Rh(PPh$_3$)$_3$Cl, 3000 h^{-1} for Ru(H)(Cl)(PPh$_3$)$_3$ and 4000 h^{-1} for Ru(COD)(PPh$_3$)$_2$PF$_6$ under essentially the same conditions. The hydrogenation is supposed to proceed through the initial addition of lanthanide hydride complex to the olefin double bond and subsequent hydrogenolysis.

An efficient asymmetric hydrogenation of 2-phenyl-1-butene has been achieved by the use of an optically active catalyst possessing chiral auxiliaries at the cyclopentadienyl ring [3].

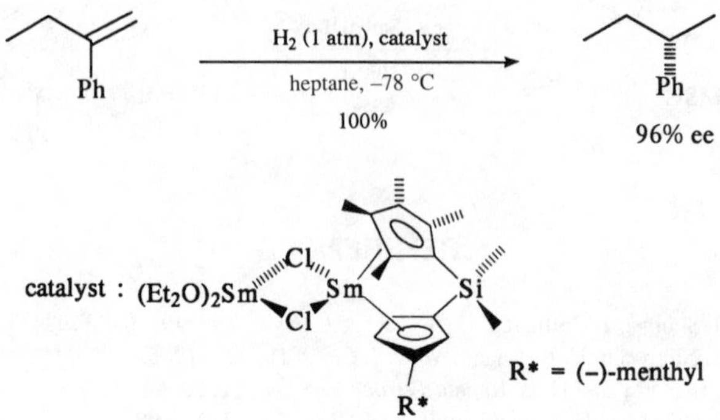

REFERENCES

1. H. Mauermann, P. N. Swepston and T. J. Marks, *Organometallics* **4**, 200 (1985).
2. G. Jeske, H. Lauke, H. Mauermann, H. Schumann and T. J. Marks, *J. Am. Chem. Soc.* **107**, 8111 (1985).
3. V. P. Conticello, L. Brard, M. A. Giardello, Y. Tsuji, M. Sabat, C. L. Stern and T. J. Marks, *J. Am. Chem. Soc.* **114**, 2761 (1992).

5.3 REDUCTIONS OF OTHER FUNCTIONAL GROUPS

Combination of existing reducing agents with lanthanide(III) salts modifies the reducing ability of the original reducing agents. Lithium aluminium hydride, when it is combined with anhydrous cerium chloride, exhibits a powerful reducing ability. Various organic halides including aliphatic and aromatic fluorides are reduced to the corresponding hydrocarbon in essentially quantitative yields by $LiAlH_4$–$CeCl_3$ in THF at reflux [1]. The same reagent system can be employed for the efficient reductions of phosphine oxide to phosphines, oximes to primary amines, and α,β-unsaturated carbonyl compounds to allylic alcohols [1–3].

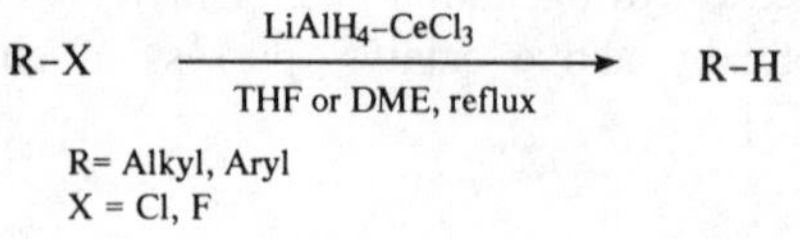

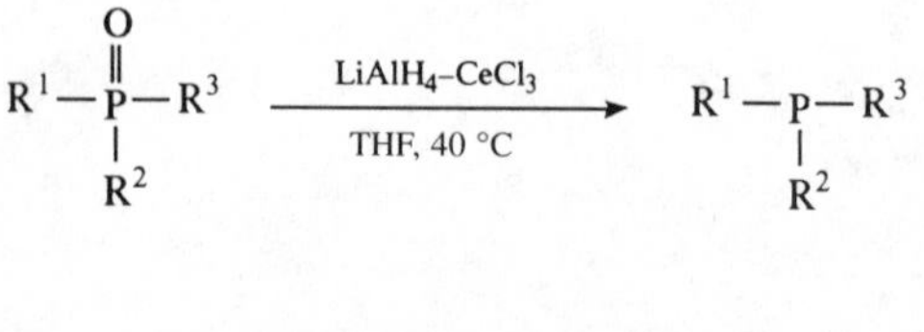

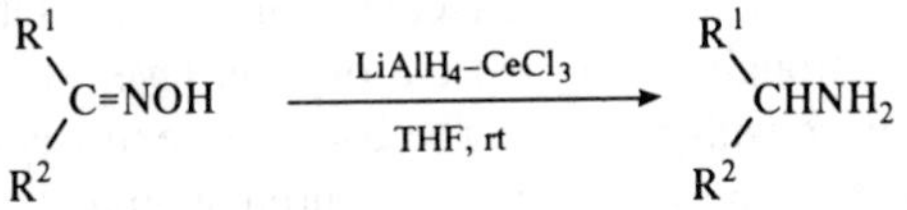

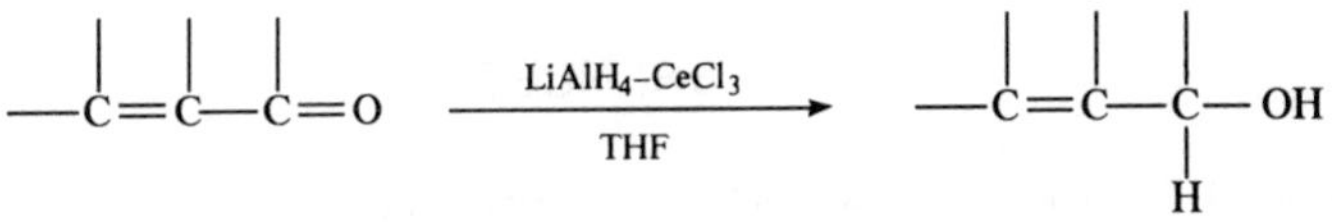

$$\begin{array}{c} R^1 \\ \diagdown \\ \diagup \\ R^2 \end{array} C=NOH \quad \xrightarrow[\text{THF, rt}]{\text{LiAlH}_4\text{-CeCl}_3} \quad \begin{array}{c} R^1 \\ \diagdown \\ \diagup \\ R^2 \end{array} CHNH_2$$

$$-\overset{|}{C}=\overset{|}{C}-\overset{|}{C}=O \quad \xrightarrow[\text{THF}]{\text{LiAlH}_4\text{-CeCl}_3} \quad -\overset{|}{C}=\overset{|}{C}-\overset{|}{\underset{H}{C}}-OH$$

Reduction of cinnamaldehyde with LiAlH$_4$ in the presence of CeCl$_3$ [3]

$$\text{PhCH=CHCHO} \quad \xrightarrow[\text{THF, rt}]{\text{LiAlH}_4\text{-CeCl}_3} \quad \text{PhCH=CHCH}_2\text{OH} \quad + \quad \text{Ph(CH}_2)_3\text{OH}$$

Cinnamaldehyde (0.132 g, 1 mmol) was added to a suspension of anhydrous cerium chloride (1 mmol) in THF (7 ml). The mixture was well stirred at room temperature for 5 min, and then LiAlH$_4$ (0.037 g, 1 mmol) was added. A vigorous gas evolution continued for 10 min before the mixture was neutralized with dilute HCl. The mixture was extracted with ether (3 x 20 ml), and the combined extracts were dried over MgSO$_4$. GC analysis of the extract revealed the presence of cinnamyl alcohol (99%) and 3-phenylpropan-1-ol (1%).

Secondary and tertiary phosphine oxides are directly converted to phosphine-boranes in high yields by treatment with LiAlH$_4$–NaBH$_4$–CeCl$_3$ [4].

$$R^1R^2R^3P(O) \quad \xrightarrow[\text{THF, rt, 2–15 h}]{\text{LiAlH}_4\text{–NaBH}_4\text{–CeCl}_3} \quad R^1R^2R^3PBH_3$$

General procedure for the preparation of phosphine-boranes from phosphine oxides [4]

To a suspension of anhydrous cerium chloride (9 mmol) in dry THF (20 ml) was added NaBH$_4$ (0.34 g, 9 mmol) under argon atmosphere, and the mixture was stirred for 1 h. The flask was immersed in an ice bath and phosphine oxide (3 mmol) was added, and then LiAlH$_4$ (0.14 g, 3.6 mmol) was added by portions. The ice bath was removed and the mixture stirred at room temperature for 2–15 h until the reaction was complete (checked by TLC). The reaction mixture was diluted with benzene (15 ml) and poured slowly to ice water containing concentrated HCl (4 ml). The mixture was filtered through Celite and the organic layer separated. The aqueous layer was extracted twice with benzene. The combined extracts were dried (Na$_2$SO$_4$) and evaporated. The residue was subjected to chromatography on a short column of silica gel to give phosphine-borane. When the product was a crystalline solid, it was recrystallized from hexane–benzene.

REFERENCES

1. T. Imamoto, T. Takeyama and T. Kusumoto, *Chem. Lett.* 1491 (1985).
2. T. Imamoto, *Rev. Heteroatom Chem.* **3**, 87 (1990).
3. S. Fukuzawa, T. Fujinami, S. Yamauchi and S. Sakai, *J. Chem. Soc. Perkin Trans. 1*, 1929 (1986).
4. T. Imamoto, T. Kusumoto, N. Suzuki and K. Sato, *J. Am. Chem. Soc.* **107**, 5301 (1985).

5.4   CARBON–CARBON BOND-FORMING REACTIONS

5.4.1   Organolanthanide(III) reagents

Organolanthanide(III) reagents can be prepared by the reaction of anhydrous lanthanide halides or triflates with organolithium reagents or Grignard reagents. These reagents have enhanced reactivities toward carbonyl groups and undergo carbonyl addition reactions even with carbonyl substrates that are prone to side reactions such as enolization and 1,4-addition. In this section, the scope of their reactivities as well as their synthetic applications are described, with the emphasis on the utility of organocerium reagents.

A full discussion of the reactivities of organolanthanide(III) reagents is beyond the scope of this book; several reviews of this subject have been published [1–5].

REFERENCES

1. G. A. Molander, in *The Chemistry of the Metal Carbon Bond*, Vol. 5, ed. F. R. Hartley, Chap. 8. John Wiley, New York, 1989.
2. G. A. Molander, *Chem. Rev.* **92**, 29 (1992).
3. G. A. Molander, in *Comprehensive Organic Synthesis*, Vol. 1, ed. B. M. Trost, I. Fleming and S. L. Schreiber, Chap. 1.9. Pergamon, London, 1991.
4. T. Imamoto, *Pure Appl. Chem.* **62**, 747 (1990).
5. T. Imamoto, in *Comprehensive Organic Synthesis*, Vol. 1, ed. B. M. Trost, I. Fleming and S. L. Schreiber, Chap. 1.8. Pergamon, London, 1991.

5.4.1.1   Oganocerium reagents

5.4.1.1.1   Organolithium–cerium halide system

Cerium has the greatest natural abundance of the rare earths, and its salts are commercially available and are the cheapest in the lanthanide series. Cerium halides such as $CeCl_3$ and CeI_3 react with organolithium reagents in THF at –78 °C to generate organocerium reagents [1,2]. Although the structures of the reagents have not yet been investigated, the reagents prepared from simple alkyllithiums with cerium halides ($CeCl_3$ or CeI_3) are supposed to involve a carbon–cerium σ-bond.

Organocerium reagents can be prepared from a variety of organolithium compounds including heteroatom stabilized organolithium compounds. The most convenient way to prepare cerium reagents is to use anhydrous cerium chloride. Although anhydrous cerium chloride is commercially available, it is readily prepared by the dehydration of cerium chloride heptahydrate *in vacuo*.

The reagents readily undergo nucleophilic addition to carbonyl compounds at low temperature; while at temperatures above 0 °C the reagents decompose and the reaction with carbonyl compounds at these temperatures affords a mixture of reduction products [1].

Characteristic features of the reactions are as follows:

(1) Organocerium reagents undergo nucleophilic addition to carbonyl groups like organolithium reagents and Grignard reagents. The addition reactions usually occur much more smoothly than the parent organolithium reagents. Various carbonyl compounds are converted to alcohols in high yields, even though the substrates are susceptible to enolization, metal–halogen exchange, reduction and pinacol coupling reaction by simple lithium reagents [1,2].

(2) Reactions with α,β-unsaturated carbonyl compounds give 1,2-addition products selectively [3,4]. Use of less sterically hindered reagents results in higher levels of selectivity, while sterically crowded reagents result in lower levels.

(3) Allylic cerium reagents tend to react at less substituted termini [5,6]. This is in sharp contrast to the reactions of other organometallic reagents such as titanium reagents.

(4) High diastereoselectivities are achieved. In some cases opposite stereo-selectivities are observed compared with the reactions of corresponding organo-lithium reagents [7,8].

(5) The reagents add to imines, but not so efficiently as to carbonyl groups [9].

(6) The reagents add to chiral hydrazones in excellent yields and high diastereoselectivity [10–13].

(7) The reagents add to nitriles twice at temperatures ranging from –65 °C to –25 °C to give tertiary carbinamines [14].

(8) For epoxides, the reagents promote ring opening [15].

Table 5.4 shows representative reactions of organocerium reagents. The reagents are useful in the synthesis of alcohols that are difficult to prepare using organolithium reagents or Grignard reagents.

Preparation of anhydrous cerium chloride [1]

Cerium chloride heptahydrate (0.56 g, 1.5 mmol) was quickly and finely ground to powder in a mortar and placed in a 30 ml two-necked flask equipped with a stirrer bar and a three-way cock. The flask was immersed in an oil bath and heated gradually to 135–140 °C with evacuation (<0.5 mmHg). After 1 h at this temperature, the cerium chloride was completely dried *in vacuo* by stirring at the same temperature for an additional hour. While the flask was still hot, argon gas was introduced and the flask was then cooled in an ice bath. Tetrahydrofuran (5 ml) freshly distilled from sodium benzophenone was added all at once with vigorous stirring. The ice bath was removed and the suspension was well stirred overnight under argon at room temperature.

The following procedure for the preparative-scale production of anhydrous cerium chloride is recommended. Cerium chloride heptahydrate (*ca.* 20 g) was quickly ground to a powder in a mortar and placed in a round-bottomed flask equipped with a three-way cock. The flask was connected to a vacuum pump through a dry ice trap and heated *in vacuo* at 100 °C for 2 h with occasional swirling. The resulting opaque solid was quickly pulverized in a mortar and replaced in the same flask. A stirrer bar was then added and the bath temperature was raised to 130–140 °C. At this temperature cerium chloride was completely dried *in vacuo* with stirring for 2–3 h.

Anhydrous cerium chloride so obtained can be stored in a sealed vessel. Cerium chloride is extremely hygroscopic, and hence it is recommended that it be dried *in vacuo* at 140 °C for 1–2 h before use.

Preparation of organocerium reagents and reactions with carbonyl compounds. General procedure [1]

$$RLi \quad + \quad CeCl_3 \quad \longrightarrow \quad RCeCl_2$$

$$\xrightarrow{\text{R}^1\text{COR}^2} \quad R-\underset{\underset{\displaystyle R^1}{|}}{\overset{\overset{\displaystyle OH}{|}}{C}}-R^2$$

Under argon atmosphere, dry THF (5 ml) was added all at once with vigorous stirring to anhydrous cerium chloride at 0 °C. The ice bath was removed, and the suspension well stirred under argon at room temperature overnight (stirring for more than 3 h is essential). The resulting milky suspension was cooled to –78 °C, and to this was added an organolithium reagent (1.5 mmol) with stirring. After 30 min, the carbonyl compound (1 mmol) was added and stirring was continued until the reaction was completed. The reaction was quenched with dilute HCl or dilute acetic acid, followed by extraction with ether. The combined extracts were dried (MgSO$_4$) and concentrated. Purification of the residue in the usual manner afforded the addition product.

Reaction of butylcerium reagent with benzonitrile; preparation of 5-amino-5-phenylnonane [14]

$$BuLi \quad + \quad CeCl_3 \quad \longrightarrow \quad BuCeCl_2$$

$$2\,BuCeCl_2 \quad + \quad PhCN \quad \longrightarrow \quad PhC(NH_2)Bu_2$$

Butyllithium (16 ml of a 2.5 M solution of hexane) was added to a suspension of anhydrous cerium chloride (40.2 mmol) in THF (80 ml), keeping the temperature below –50 °C. The mixture was stirred in a dry ice–acetone bath for 30 min, and benzonitrile (1.34 g, 13 mmol) in 2 ml of THF was added. Stirring at –65 °C was continued for 5 h. Concentrated NH_4OH (25 ml) was added at less than –40 °C, and the mixture was brought to 25 °C and filtered using Celite. The solids were washed several times with CH_2Cl_2, and the aqueous layer of the filtrates extracted twice with CH_2Cl_2. The combined organic phases were dried and concentrated. The residue was taken up in 30 ml of toluene and stirred with 30 ml of 3% H_3PO_4 for 15 min. The toluene layer was extracted with two 10-ml portions of water, and the combined aqueous phases were washed once with toluene and made basic with concentrated NH_4OH. The mixture was extracted several times with CH_2Cl_2, and the residue obtained on removal of the solvent from the dried ($MgSO_4$) extracts was short-path distilled to give 5-amino-5-phenylnonane (2.53 g, 90%), boiling at a bath temperature of 80–130 °C (0.003 mmHg). The hydrochloride of 5-amino-5-phenylnonane has a melting point of 209–216 °C after crystallization from water.

REFERENCES

1. T. Imamoto, T. Kusumoto, Y. Tawarayama, Y. Sugiura, T. Mita, Y. Hatanaka and M. Yokoyama, *J. Org. Chem.* **49**, 3904 (1984).
2. T. Imamoto, Y. Sugiura and N. Takiyama, *Tetrahedron Lett.* **25**, 4233 (1984).
3. T. Imamoto and Y. Sugiura, *J. Organometal. Chem.* **285**, C21 (1985).
4. T. Imamoto and Y. Sugiura, *J. Phys, Org. Chem.* **2**, 93 (1989).
5. B.-S. Guo, W. Doubleday and T. Cohen, *J. Am. Chem. Soc.* **109**, 4710 (1987).
6. T. Cohen and M. Bhupathy, *Acc. Chem. Res.* **22**, 152 (1989).
7. M. Kawasaki, F. Matsuda and S. Terashima, *Tetrahedron Lett.* **26**, 2693 (1985).
8. T. Izawa, Y. Nishimura and S. Kondo, *Carbohydrate Res.* **211**, 137 (1991).
9. M. Wada, S. Nishimura and T. Imamoto, unpublished results.
10. S. E. Denmark, T. Weber and D. W. Piotrowski, *J. Am. Chem. Soc.* **109**, 2224 (1987).
11. S. E. Denmark, J. P. Edwards and O. Nicaise, *J. Org. Chem.* **58**, 569 (1993).
12. S. E. Denmark and O. Nicaise, *Synlett* 359 (1993).
13. D. Enders, R. Funk, M. Klatt, G. Raabe and E. R. Hovestreydt, *Angew. Chem. Int. Ed. Engl.* **32**, 418 (1993).

TABLE 5.4

Reactions of Organocerium Reagents

Reagents	Carbonyl compound	Product (yield %)	Ref.
Alkylcerium reagent			
$CH_3Li–CeI_3$	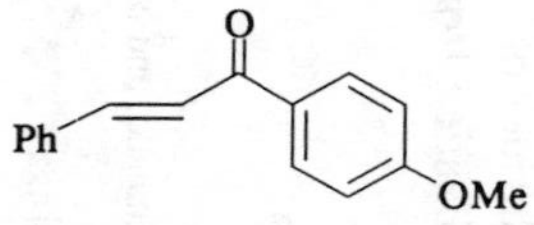 (4-OMe-phenyl)–CO–CH=CH–Ph	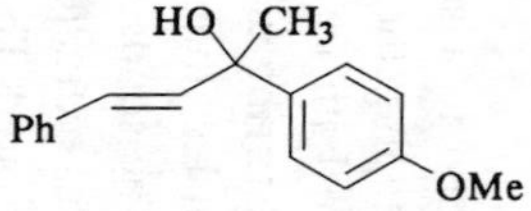 (4-OMe-phenyl)–C(OH)(CH_3)–CH=CH–Ph (98) (65)[a]	[3]
$CH_3Li–CeI_3$	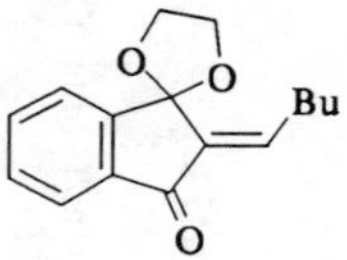 dioxolane-indanone with Bu=CH substituent	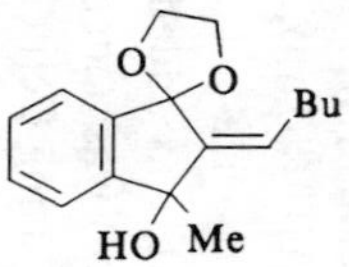 (98)	[16]
$CH_3Li–CeCl_3$	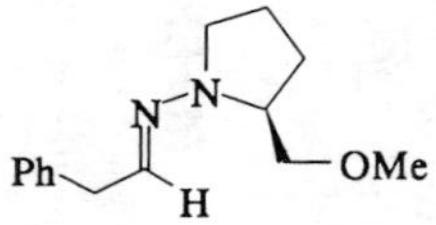 pyrrolidine–CH=N–CH$_2$Ph with OMe	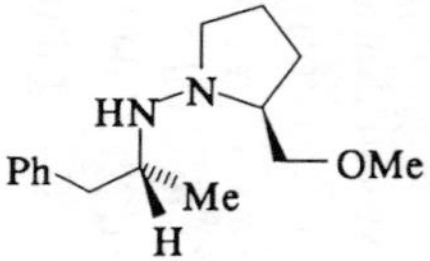 pyrrolidine–CH(Me)–NH–CH(CH$_2$Ph), OMe (>66, 92% de) (0)[a]	[10]
$BuLi–CeCl_3$	$(PhCH_2)_2CO$	$(PhCH_2)_2C(OH)Bu$ (96) (33)[a]	[1]
$BuLi–CeCl_3$[b]	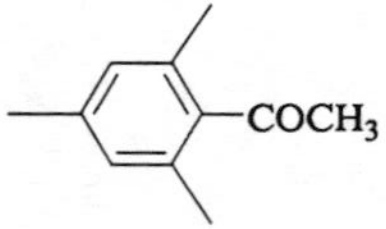 mesityl–COCH$_3$	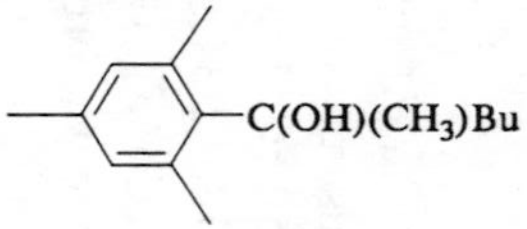 mesityl–C(OH)(CH$_3$)Bu (57) (<10)	[1]

Continued

Reagent	Substrate	Product (yield)	Ref.
BuLi–CeI₃	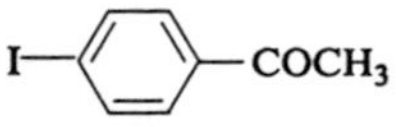 4-I–C₆H₄–COCH₃	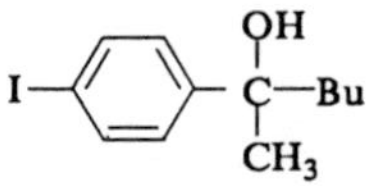 I–C₆H₄–C(OH)(Bu)(CH₃) (93) (trace)	[1]
BuLi–CeCl₃	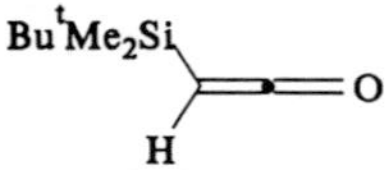 Buᵗ Me₂Si–CH=C=O	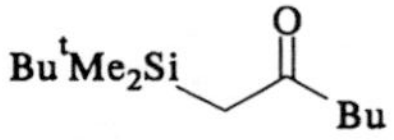 Buᵗ Me₂Si–CH₂–CO–Bu (91)	[17]
BuLi–CeCl₃	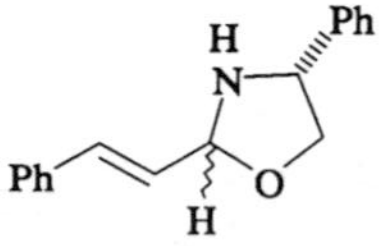	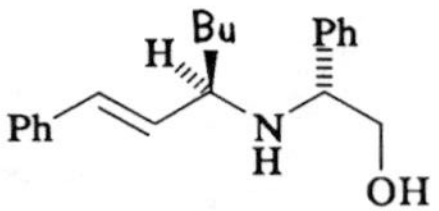 (63, 96% de) (40, 64% de)ᵃ	[18]
BuLi–CeCl₃ᵇ	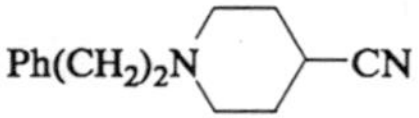 Ph(CH₂)₂N-piperidine-4-CN	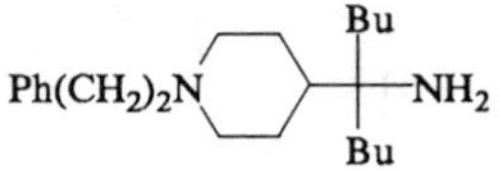 Ph(CH₂)₂N-piperidine-4-C(NH₂)(Bu)(Bu) (90)	[14]

TABLE 5.4 Continued

Reagents	Carbonyl compound	Product (yield %)	Ref.
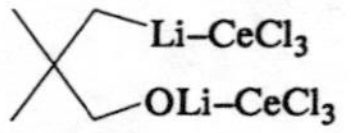 Li–CeCl₃ OLi–CeCl₃ (neopentyl structure)	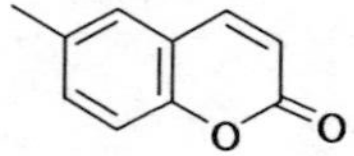 6-methylcoumarin structure (O=C–O)	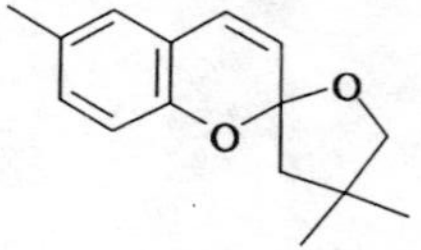 (76) spiro chromene structure	[19]
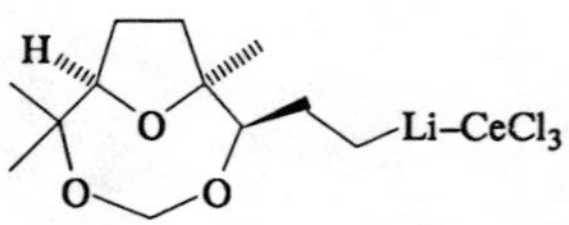 Li–CeCl₃ bicyclic ether structure	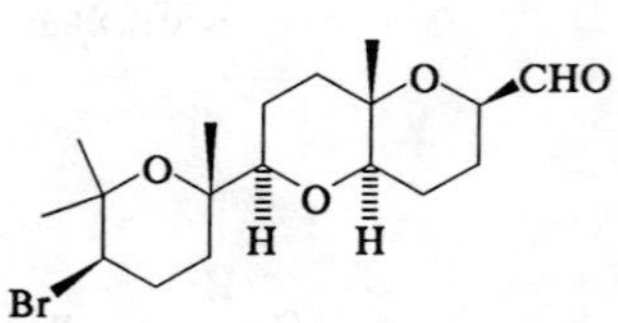 CHO bicyclic pyran structure with Br	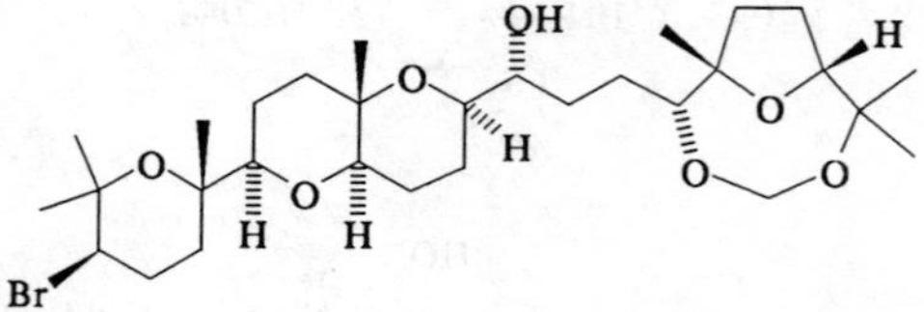 (85) complex polyether structure with Br	[20]
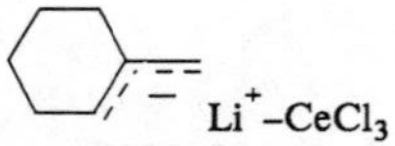 Li⁺–CeCl₃ cyclohexenyl allenyl structure	MeCH=CHCHO	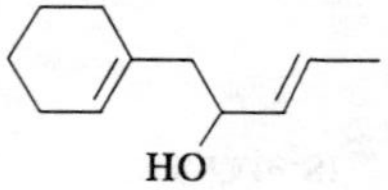 (82) cyclohexenylmethyl allylic alcohol (HO) structure	[6]
[(H₂C=CHCH₂)₄Ce]⁻ Li⁺	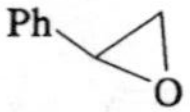 Ph epoxide (styrene oxide)	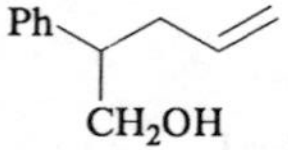 Ph OH (82) 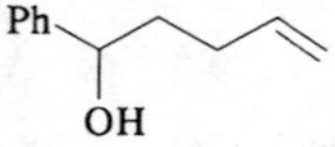 + Ph CH₂OH (15)	[15]

Alkenylcerium reagent			
CH_2=C(CH_3)Li–$CeCl_3$	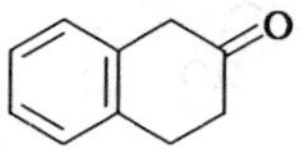	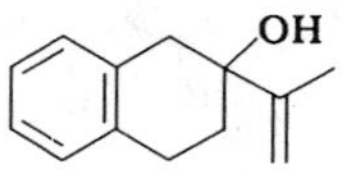 (88) (12)[a]	[2]
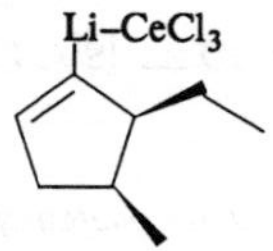	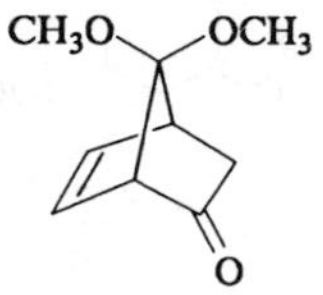	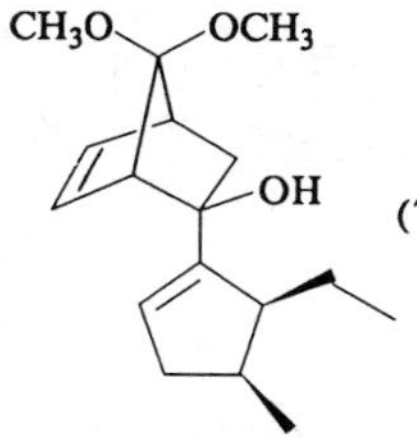 **(76)**	[21]
Arylcerium reagent			
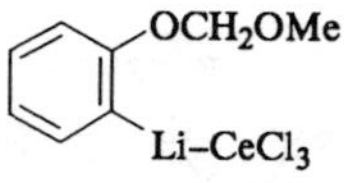	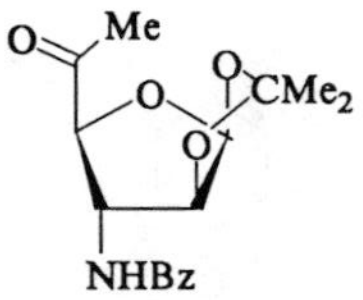	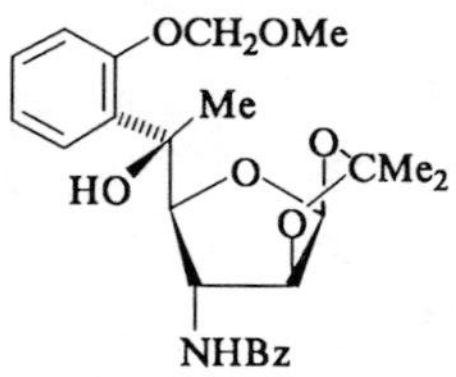 (85) (14)[a] 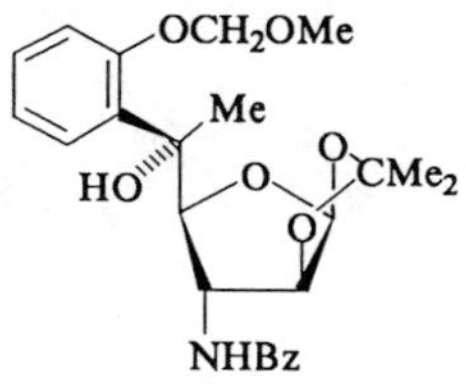 (10) (14)[a]	[8]

Continued

TABLE 5.4 Continued

Reagents	Carbonyl compound	Product (yield %)	Ref.

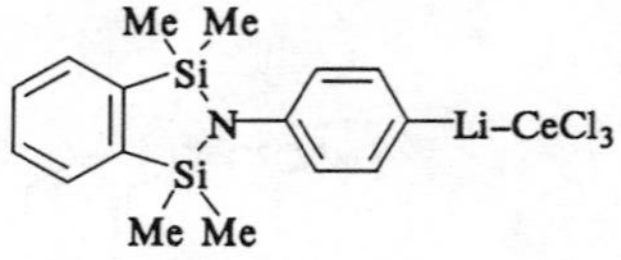

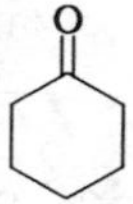

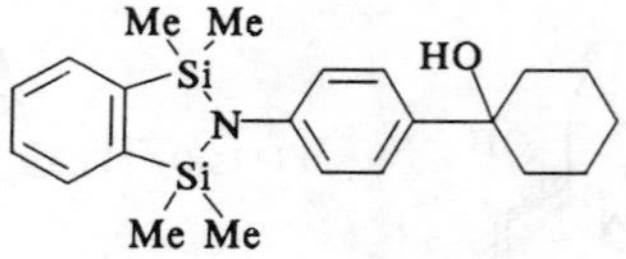

Reagents column:

Me Me
Si—N—Si
Me Me
(with benzene rings and 4-Li–CeCl$_3$ phenyl)

Br, OCH$_2$OMe, Li–CeCl$_3$ substituted arene

Alkynylcerium reagent

Me$_3$SiC≡CLi–CeCl$_3$

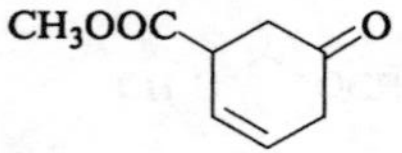

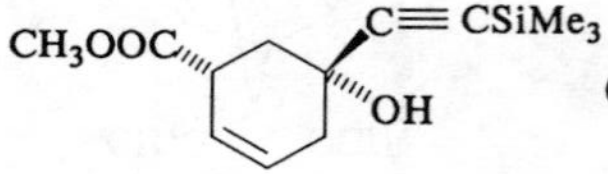

Me$_3$SiC≡CLi–CeCl$_3$

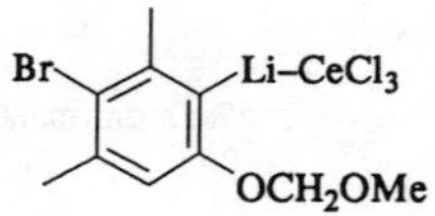

Carbonyl compound column:

cyclohexanone (O)

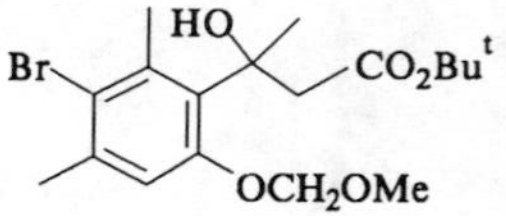

CH$_3$COCH$_2$CO$_2$Bu^t

CH$_3$OOC— (cyclohexenone)

naphthacenequinone structure with OH, OH, O, O

Product (yield %) column:

Me Me Si—N—Si Me Me, HO, cyclohexyl-phenyl (95)

HO, CO$_2$Bu^t, Br, OCH$_2$OMe (>78)

C≡CSiMe$_3$, OH, CH$_3$OOC (95)

SiMe$_3$, OH, OH, OH, O, O (77)

Ref. column:

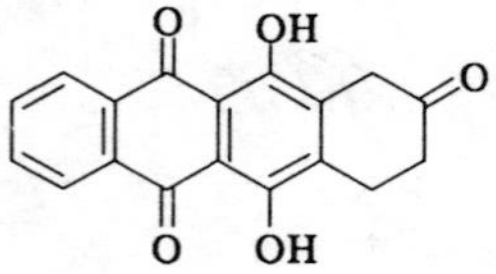

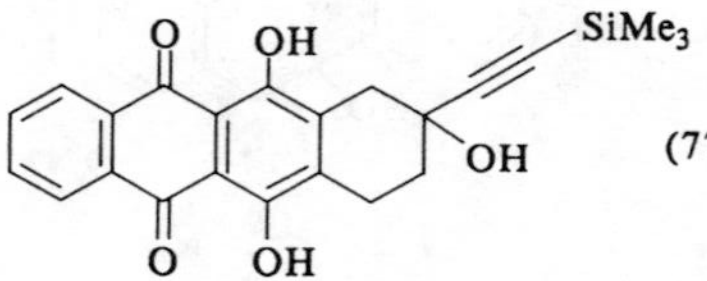

[22]

[23]

[24]

[25]

PhC≡CLi-CeCl₃	Br—C₆H₄—COCH₂Br	![structure] OH, (4-BrC₆H₄)C(CH₂Br)C≡CPh (95) (trace)[a]	[2]

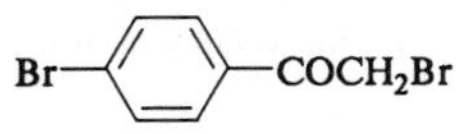

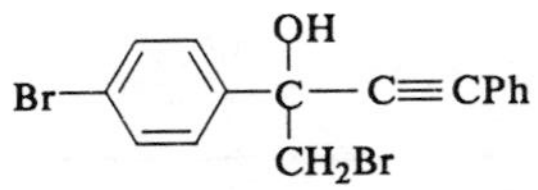

$PhC{\equiv}CLi\text{-}CeCl_3$

$Br{-}\!\!\!\bigcirc\!\!\!{-}COCH_2Br$

(95) (trace)[a] [2]

$EtOCOC{\equiv}CLi\text{-}CeCl_3$

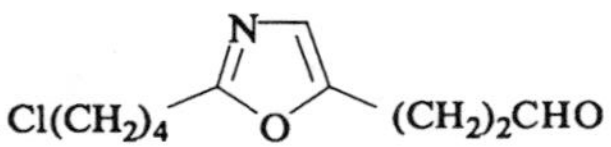

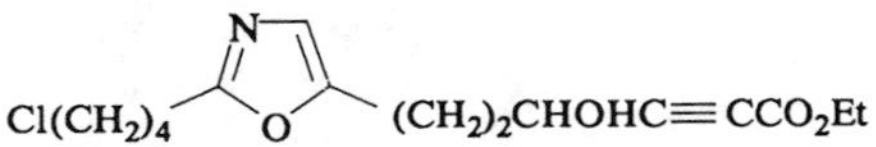

$Cl(CH_2)_4$ [oxazole] CHO

$Cl(CH_2)_4$ [oxazole] $(CH_2)_2CHOHC{\equiv}CCO_2Et$ (95) [26]

α-Heterosubstituted organocerium reagent

$Me_3SiCH_2Li\text{-}CeCl_3$

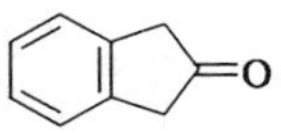

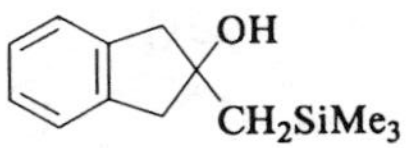

[indan-2-one]

[2-hydroxy-2-(trimethylsilylmethyl)indane: OH, CH₂SiMe₃] (83) [27]

$Me_3SiCH_2Li\text{-}CeCl_3$ (2 equiv.)

$PhCH{=}CHCOCl$

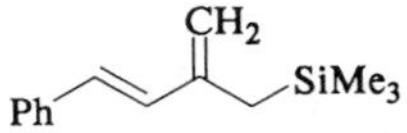

CH_2, $SiMe_3$, Ph (90) (40)[a] [28]

$Cl_2HCLi\text{-}CeCl_3$

[SESHN cyclohexane fused dioxolane, Me, Me, ketone, OH, Me, H, H]

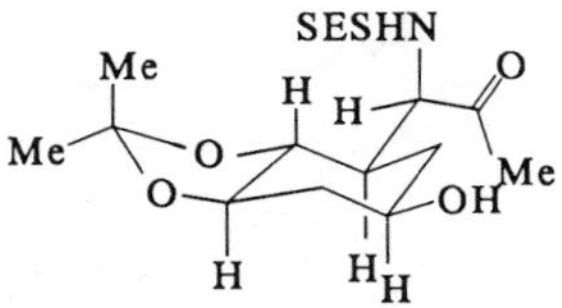

[product: SESHN, CHCl₂, OH, Me, OH, H, H, dioxolane Me Me] (54) [29]

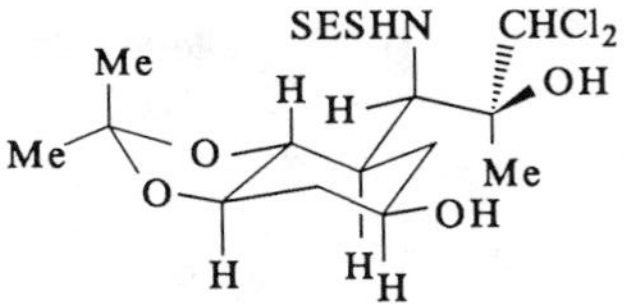

SES = $Me_3SiCH_2CH_2SO_2$-

Continued

TABLE 5.4 *Continued*

Reagents	Carbonyl compound	Product (yield %)	Ref.
NCCH$_2$Li–CeCl$_3$	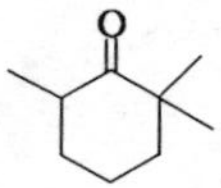 (2,2,6-trimethylcyclohexanone)	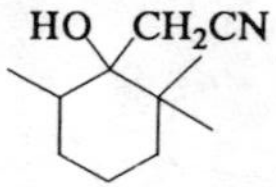 HO CH$_2$CN (88) (40)[a]	[30]
PhCH(CN)Li–CeCl$_3$	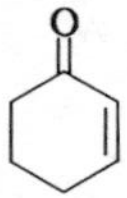 (cyclohex-2-enone)	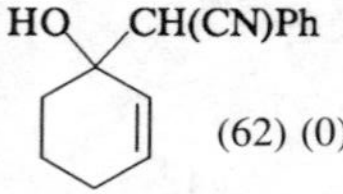 HO CH(CN)Ph (62) (0)[a] 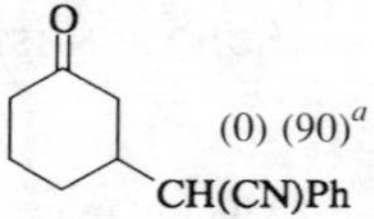 ; CH(CN)Ph (0) (90)[a]	[4]
(MeS)$_3$CLi–CeCl$_3$	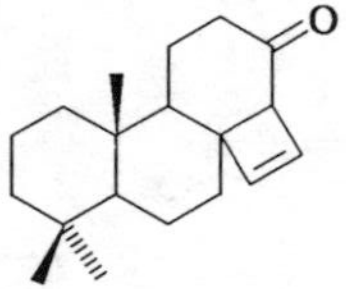 (octahydronaphthalenone)	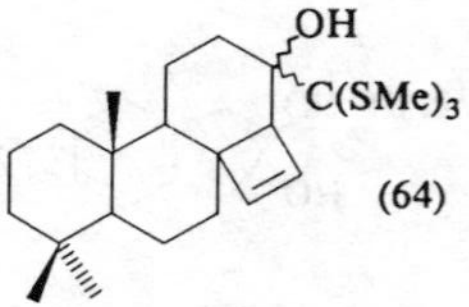 OH C(SMe)$_3$ (64)	[31]
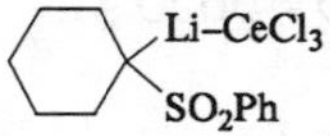 Li–CeCl$_3$ SO$_2$Ph (cyclohexyl)	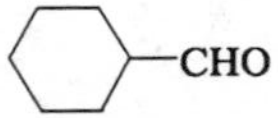 CHO (cyclohexane)	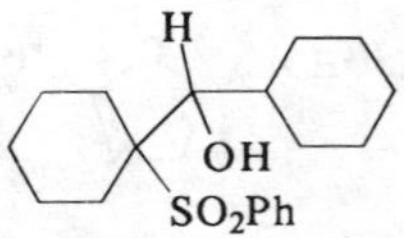 H OH OH SO$_2$Ph (100) (75)[a]	[32]

[a] Yield obtained by the use of organolithium reagent alone.
[b] Three equivalents of the reagent are employed.

14. E. Ciganek, *J. Org. Chem.* **57**, 4521 (1992).
15. S. Fukuzawa and S. Sakai, *Bull. Chem. Soc. Jpn* **65**, 3308 (1992).
16. D. Mitchell and L. S. Liebeskind, *J. Am. Chem. Soc.* **112**, 291 (1990).
17. Y. Kita, S. Matsuda, S. Kitagaki, Y. Tsuzuki and S. Akai, *Synlett* 401 (1991).
18. L. N. Pridgen, M. K. Mokhallalati and M.-J. Wu, *J. Org. Chem.* **57**, 1237 (1992).
19. B. Mudryk, C. A. Shook and T. Cohen, *J. Am. Chem. Soc.* **112**, 6389 (1990).
20. E. J. Corey and D.-C. Ha, *Tetrahedron Lett.* **29**, 3171 (1988).
21. L. A. Paquette, J. L. Romine, H.-S. Lin and J. Wright, *J. Am. Chem. Soc.* **112**, 9284 (1990).
22. R. P. Bonar-Law, A. P. Davis and B. J. Dorgan, *Tetrahedron Lett.* **31**, 6721 (1990).
23. K. Nagasawa and K. Ito, *Heterocycles* **28**, 703 (1989).
24. T. Chamberlain, X. Fu, J. T. Pechacek, X. Peng, D. M. S. Wheeler and M. M. Wheeler, *Tetrahedron Lett.* **32**, 1707 (1991).
25. Y. Tamura, M. Sasho, S. Akai, H. Kishimoto, J. Sekihachi and Y. Kita, *Chem. Pharm. Bull.* **35**, 405 (1987).
26. E. Vedejs and D. W. Piotrowski, *J. Org. Chem.* **58**, 1341 (1993).
27. C. R. Johnson and B. D. Tait, *J. Org. Chem.* **52**, 281 (1987).
28. M. B. Anderson and P. L. Fuchs, *Synth. Commun.* **17**, 621 (1987).
29. R. S. Garigipati, D. M. Tschaen and S. M. Weinreb, *J. Am. Chem. Soc.* **112**, 3475 (1990).
30. H.-J. Liu and N. H. Al-said, *Tetrahedron Lett.* **32**, 5473 (1991).
31. A. Abad, M. Arnó, M. L. Marín and R. J. Zaragozá, *Synlett* 789 (1991).
32. M. B. Anderson, M. Lamothe and P. L. Fuchs, *Tetrahedron Lett.* **32**, 4457 (1991).

5.4.1.1.2   *Grignard reagent–cerium chloride system*

Reactions of Grignard reagents with carbonyl substrates provide one of the most important methods for carbon–carbon bond formation. However, the reactions usually accompany so-called abnormal reactions (enolization, reduction, condensation, conjugate addition, pinacol coupling), and in some cases such reactions prevail over the normal reactions. The use of cerium chloride as an additive significantly enhances normal addition [1]. The Grignard reagent–cerium chloride system is useful and has widespread synthetic applicability. Table 5.5 shows various examples of reactions conducted using the Grignard reagent–cerium chloride system.

The characteristic features of this method are as follows:

(1) Abnormal Grignard reactions are remarkably suppressed, and normal addition products are produced in high yields.
(2) The reactions can be carried out at 0 °C to room temperature except for the reaction of alkenyl Grignard reagents. Alkenyl Grignard reagents react spontaneously with cerium chloride at temperatures higher than 0 °C to give dimeric and trimeric substrates [10]; these reactions should therefore be carried out at −78 °C.
(3) Good to high diastereoselectivities are observed. In some cases, the selectivities are different from those obtained by the use of Grignard reagents alone.

TABLE 5.5
Reactions of Grignard Reagents with Carbonyl Compounds or Related Substrates in the Presence of Cerium Chloride.

Reagents	Carbonyl compound	Product (yield %)	Ref.
$CH_3MgBr–CeCl_3$	$(C_2H_5)_3CCOCH_3$	$(C_2H_5)_3CC(OH)(CH_3)_2$ (95) (0)[a]	[1]
$CH_3MgBr–CeCl_3$	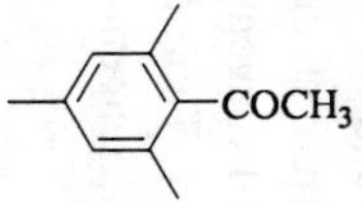	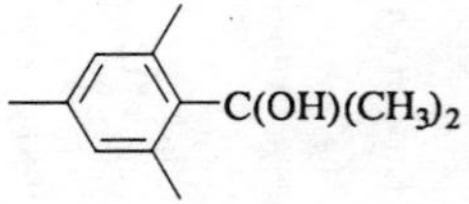 (>85) (trace)[a]	[2]
$Pr^iMgCl–CeCl_3$	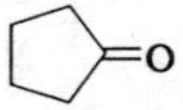	 (72) (3)[a]	[1]
 CH_2MgBr	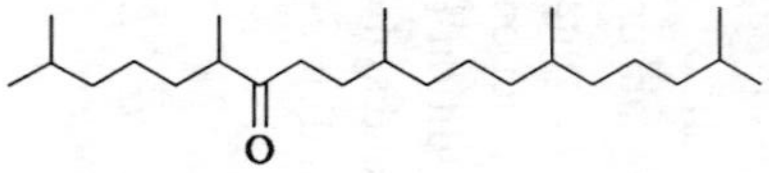	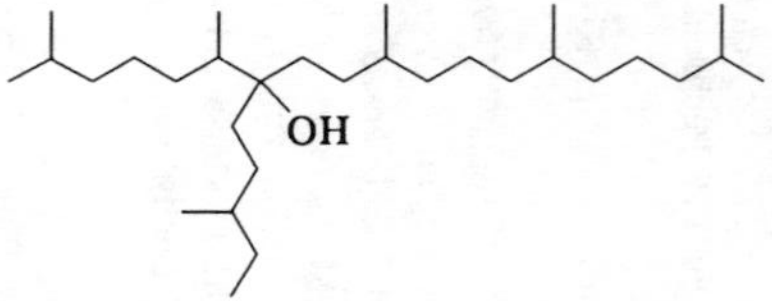 (76) (2)[a]	[3]

Continued

Reagent	Substrate	Product	Ref.
Me₂C=CH(CH₂)₂MgBr–CeCl₃		~(CH₂)₂CH=CMe₂ (72) (0)[a]	[1]
CH₂=CHMgBr–CeCl₃		(95)	[4]
CH₂=CMeMgBr–CeCl₃		(95)	[5]
PhMgBr–CeCl₃	PhCH=CHCOPh	PhCH=CHC(OH)Ph₂ (89) (5)[a] Ph₂CHCH₂COPh (11) (81)[a]	[1]

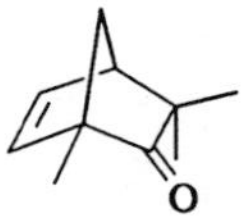

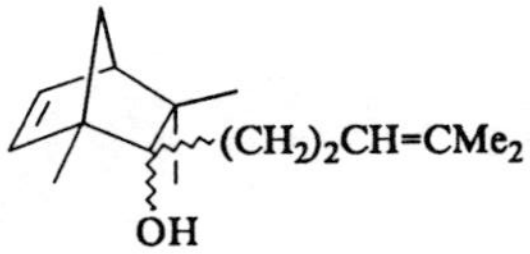

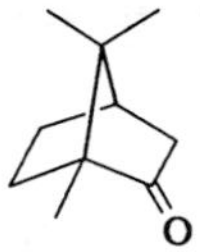

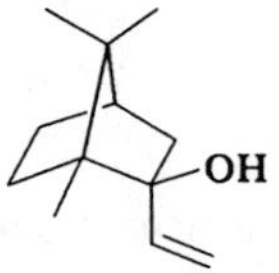

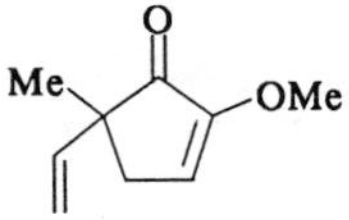

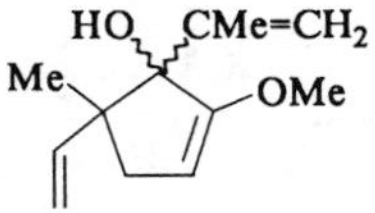

TABLE 5.5 *Continued*

Reagents	Carbonyl compound	Product (yield %)	Ref.
 $-CH_2MgBr-CeCl_3$	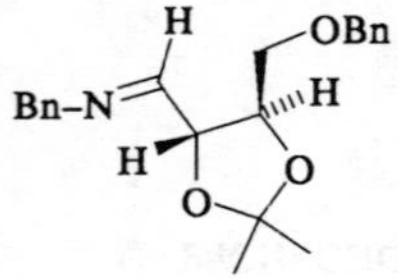	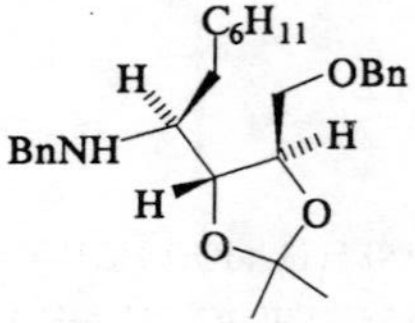 (75)	[6]
$PhMgBr-CeCl_3$	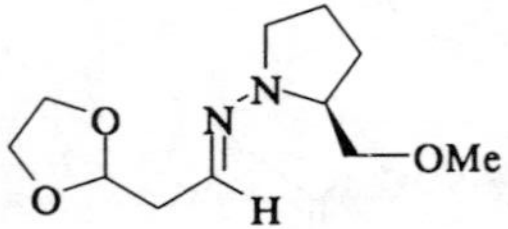	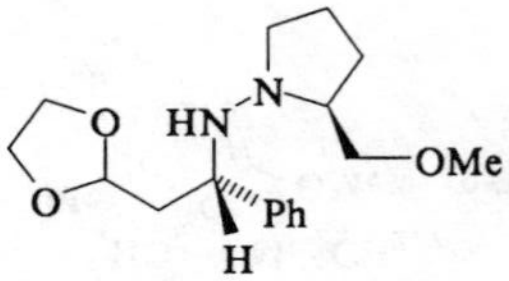 (97, 98.5% ee)	[7]
Pr^iMgCl (2 equiv.)	$PhCH_2CO_2Me$	$PhCH_2C(OH)Pr^i_2$ (97) (0)[a]	[1]
$Me_3SiCH_2MgCl-CeCl_3$	$(MeO)_2CH(CH_2)_2CO_2Me$	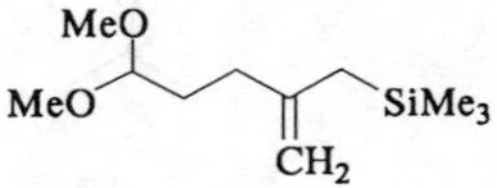 (78)	[8]

BuMgBr–CeCl$_3$ PhCH$_2$CONMe$_2$ [1]

PrMgCl–CeCl$_3$ PhCH$_2$COBu (66) (8)^a (66)   [9]

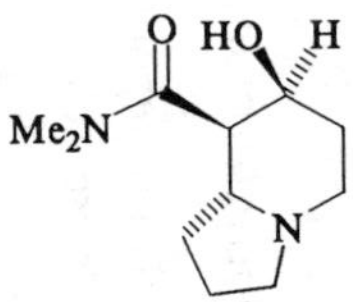

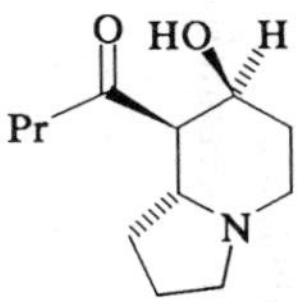

^{a}Yield obtained by the use of the Grignard reagent alone.

Reaction of a Grignard reagent with a carbonyl compound in the presence of trs to pu 96 cerium chloride. General procedure [1]

Under argon atmosphere, dry THF (5 ml) was added all at once with vigorous stirring to anhydrous cerium chloride at 0 °C. The ice bath was removed, and the suspension well stirred under argon at room temperature overnight. The flask was again immersed in an ice bath and the Grignard reagent (1.5 mmol) added by syringe. After the mixture had been stirred for 1.5 h at 0 °C, a solution of the carbonyl compound (1 mmol) in THF (1–2 ml) was added and stirring continued for 30 min. The reaction mixture was treated with 10% aqueous acetic acid (10 ml). The product was extracted into ether, and the combined extracts washed successively with brine, aqueous $NaHCO_3$ and brine, and dried over $MgSO_4$. The solvent was evaporated and the residue subjected to preparative TLC to give the addition product.

REFERENCES

1. T. Imamoto, N. Takiyama, K. Nakamura, T. Hatajima and Y. Kamiya, *J. Am. Chem. Soc.* **111**, 4392 (1989).
2. R. F. Evilia, D. Pan, J. W. Timberlake and S. L. Whittenburg, *Tetrahedron Lett.* **32**, 871 (1991).
3. J. N. Robson and S. J. Rowland, *Tetrahedron Lett.* **29**, 3837 (1988).
4. D. S. Keegan, M. M. Midland, R. T. Werley and J. I. McLoughlin, *J. Org. Chem.* **56**, 1185 (1991).
5. J. W. Herndon, L. A. McMullen and C. E. Daitch, *Tetrahedron Lett.* **31**, 4547 (1990).
6. T. Matsumoto, Y. Kobayashi, Y. Takemoto, Y. Ito, T. Kamijo, H. Harada and S. Terashima, *Tetrahedron Lett.* **31**, 4175 (1990).
7. D. Enders, M. Klatt and R. Funk, *Synlett* 226 (1993).
8. T. V. Lee, J. A. Channon, C. Cregg, J. R. Porter, F. S. Roden and H. T.-L. Yeoh, *Tetrahedron* **45**, 5877 (1989).
9. D. L. Comins and H. Hong, *J. Am. Chem. Soc.* **113**, 6672 (1991).
10. T. Imamoto, T. Hatajima and K. Ogata, *Tetrahedron Lett.* **32**, 2787 (1991).

5.4.1.2 Organoytterbium reagents

Although relatively few reports have appeared on organoytterbium(III) reagents, they show interesting stereoselectivities, as exemplified in the following scheme. The enhanced diastereoselectivities are ascribed to the relatively small ionic radius of ytterbium [1].

RM–Yb(OTf)$_3$ +

RM = MeLi, BuLi, Bu^{t}Li, PhLi,

C$_4$H$_9$CH=CHLi, CH$_2$=CHMgBr

>97 : <3

Interesting chelation control effects are observed in the reaction of chiral 2-acyl-1,3-oxathiane derivative [2]. The following scheme is illustrative. The stereochemical outcome of the reaction of the ytterbium reagent is reasonably explained in terms of the high coordination ability of ytterbium(III) towards not only oxygen but also suphur functionalities.

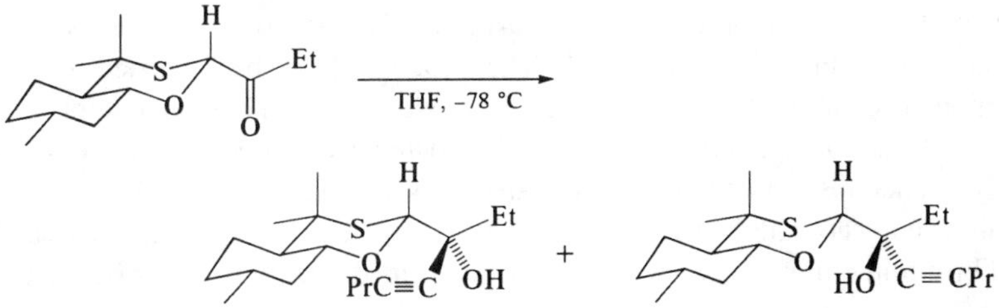

PrC≡CLi :   75 : 25   (94% yield)

PrC≡CLi –YbCl₃ :   2 : 98   (99% yield)

REFERENCES

1. G. A. Molander, E. R. Burkhardt and P. Weinig, *J. Org. Chem.* **55,** 4990 (1990).
2. K. Utimoto, A. Nakamura and S. Matsubara, *J. Am. Chem. Soc.* **112,** 8189 (1990).

5.4.2   Aldol reactions promoted by lanthanide(III) reagents

5.4.2.1   Stoichiometric reactions

Lithium enolates are converted to cerium enolates on treatment with anhydrous cerium chloride at –78 °C. The cerium enolates undergo aldol addition with a variety of carbonyl compounds [1]. The yields of the products are usually higher than the yields using lithium enolates alone. The cerium chloride-promoted aldol reactions are applicable to sterically hindered and/or easily enolizable carbonyl substrates [1,2]. The stereoselectivity of the reactions is almost the same as the cases using lithium enolates, since the reaction proceeds through a six-membered transition state.

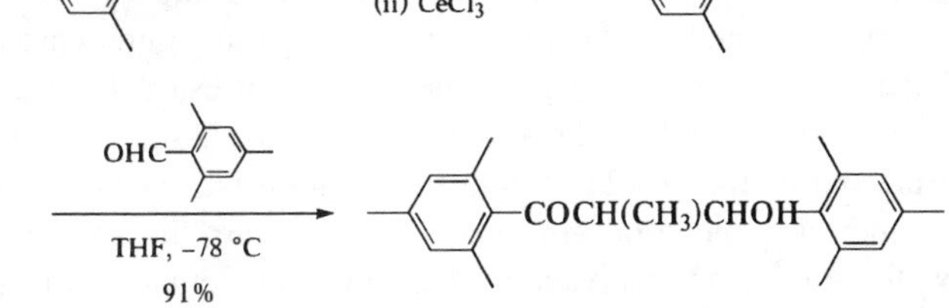

anti : *syn* = 93 : 7

Reductive coupling of α-haloketones with carbonyl compounds is remarkably promoted by lanthanide(III) ion. Use of CeI_3 as a coupling agent affords α,β-unsaturated carbonyl compounds, while the $CeCl_3$–NaI or $CeCl_3$–$SnCl_2$ reagent system leads to aldols with a mixture of *anti* and *syn* isomers [3].

Electrochemical aldol reactions of α-bromo ketones with aldehydes are also promoted by lanthanide ion [4,5]. The reactions are best carried out using stoichiometric amounts of lanthanum bromide in dry THF containing lithium perchlorate at room temperature. The electrochemical reduction of the α-bromo ketones generates (Z)-enolates in high stereoselectivity. The *syn* aldols are formed initially, but the aldol reaction is reversible under these conditions and the equilibrium mixture of *syn* and *anti* aldols is formed, the *anti* isomer being a major product.

$$R^1COCHBrR^2 \quad + \quad R^3CHO \xrightarrow[\text{LaBr}_3,\ \text{THF–LiClO}_4,\ \text{rt}]{2\,e^-} R^1COCH(R^2)CHOHR^3$$

Preparation of 4,5,7-trimethylcoumarin [2]

To a suspension of anhydrous cerium chloride (1.5 g, 6 mmol) in dry THF (12 ml) was added a solution of $LiCH_2CO_2Bu^t$(0.61 g, 5 mmol) in THF (5 ml) at −78 °C. After this had been stirred for 1.5 h, a solution of 2-methoxymethoxy-4,6-dimethylacetophenone (0.42 g, 2 mmol) in THF (5 ml) was added and the resulting mixture stirred for an additional 4 h at the same temperature. The reaction was quenched with saturated NH_4Cl (25 ml). Isolation of the organic layer by a centrifuge followed by evaporation and purification using short-path column chromatography on silica gel (pentane–ether, 3 : 1) gave the β-hydroxy ester (0.60 g, 93%) as a colourless oil. This compound (0.150 g, 0.46 mmol) was dissolved in a mixture of 20% HCl (1 ml) and methanol (2 ml), and the solution was heated at 80 °C with stirring, whereupon a white solid appeared. After 10 min, the solid material was collected by filtration and recrystallized from 95% ethanol to give 4,5,7-trimethylcoumarin (0.085 g, 98%), m.p. 182.0–182.5 °C.

REFERENCES

1.   T. Imamoto, T. Kusumoto and M. Yokoyama, *Tetrahedron Lett.* **24**, 5233 (1983).
2.   K. Nagasawa, H. Kanbara, K. Matsushita and K. Ito, *Tetrahedron Lett.* **26**, 6477 (1985).
3.   S. Fukuzawa, T. Tsuruta, T. Fujinami and S. Sakai, *J. Chem. Soc. Perkin Trans. 1*, **987**, 1473 (1987).
4.   A. J. Fry, M. Susla and M. Weltz, *J. Org. Chem.* **52**, 2496 (1987).
5.   A. J. Fry and M. Susla, *J. Am. Chem. Soc.* **111**, 3225 (1989).

5.4.2.2   Catalytic reactions

Aldol reactions of enol silyl ethers or ketene silyl acetals with carbonyl compounds are promoted by Lewis acids. Lanthanide(III) complexes are also effective as catalysts for this class of aldol reactions [1,2]. Lanthanide catalysts have some advantages over the use of rather strong Lewis acids such as $TiCl_4$, $BF_3 \cdot OEt_2$ and $SnCl_4$ in chemo- and stereoselectivities [3–7].

The following scheme represents aldehyde discrimination, where 4-nitrobenzaldehyde is less reactive than benzaldehyde [4]. This reverse selectivity is ascribed to preferential activation of benzaldehyde over 4-nitrobenzaldehyde by the europium catalyst.

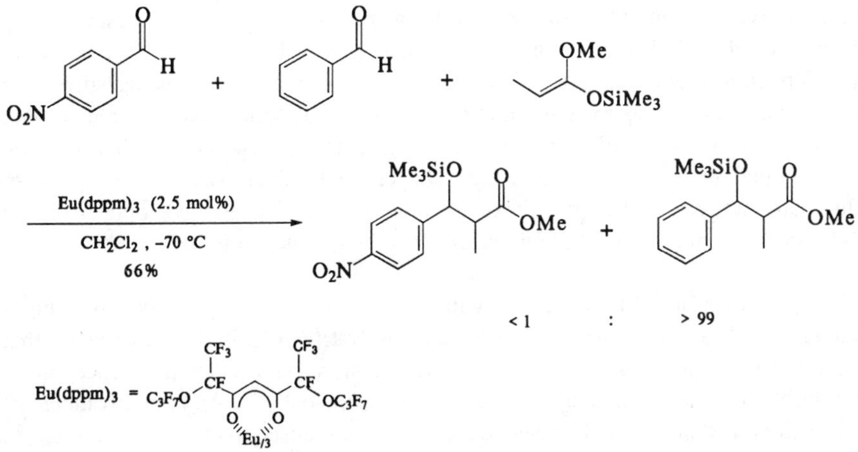

Remarkably high stereoselectivities can be achieved by the use of Ln(III) complexes [8,9]. The strong oxophilicity and high coordination ability of lanthanide(III) again effect high and unique stereoselectivities in aldol reactions. This stereoselective reaction provides a protocol for the synthesis of sugars and related substrates.

Preparation of optically pure 2-deoxy-D-ribonolactone [8]

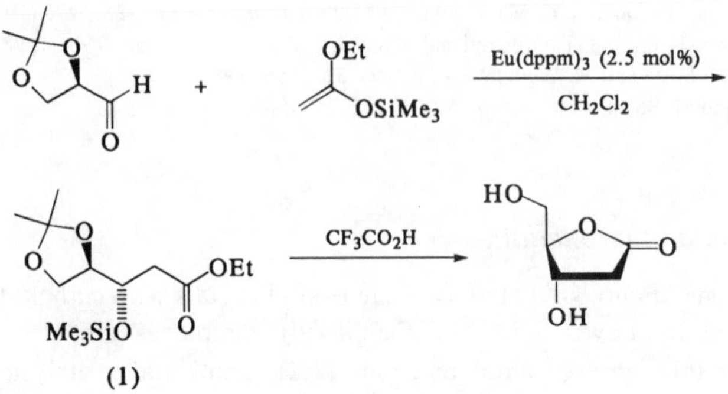

(1)

D-Glyceraldehyde acetonide (0.130 g, 1 mmol) and 1-ethoxy-1-trimethylsiloxyethene (0.240 g, 1.5 mmol) were dissolved in CH_2Cl_2 (4 ml) and the solution cooled to –78 °C. To this solution was added a solution of (–)-Eu(dppm)$_3$ (0.015 mmol) in CH_2Cl_2, and the mixture was stirred for 2 h at the same temperature. The reaction mixture was treated with saturated aqueous NH_4Cl solution and extracted with ethyl acetate (3 x 20 ml). The combined extracts were washed with brine and dried over $MgSO_4$. Evaporation of the solvent afforded a crude product (1) with an isomer ratio of *anti:syn* of 95:5. The crude product was dissolved in THF (2 ml), and to this were added trifluoroacetic acid (0.5 ml) and one drop of water. After being stirred for 1 day, the mixture was treated with aqueous $NaHCO_3$ solution and extracted with ethyl acetate (5 x 30 ml). The combined extracts were washed with brine and concentrated under reduced pressure. The crude product was purified by flash chromatography on silica gel (hexane–ethyl acetate, 5:1) to give optically pure 2-deoxy-D-ribonolactone (0.118 g, 90%), $[\alpha]_D^{20}-5.3°$ (*c* 1.2 in EtOH).

Aldol reactions of enol silyl ethers with aldehydes can be carried out by using a lanthanide trifluoromethanesulphonate as a catalyst [8–10]. It should be noted that these reactions proceed even in an aqueous medium, and hence this procedure is applicable to water-soluble aldehydes such as formaldehyde. Another advantage of this method is that the catalyst can be recovered and reused in this reaction.

Reaction of an enol trimethylsilyl ether with formaldehyde. General procedure [8]

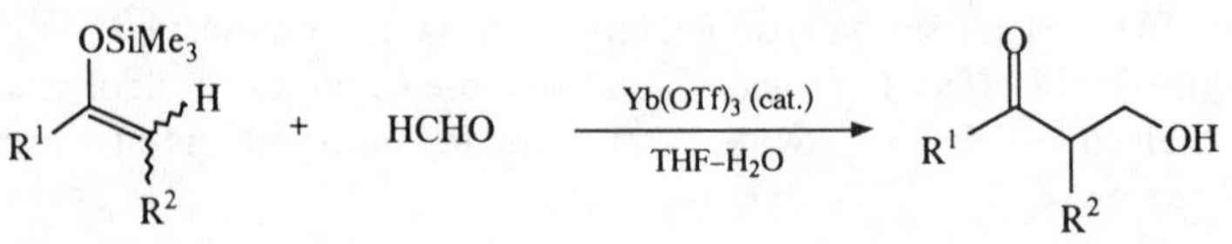

To formaldehyde (1 ml of 37% solution in water) and THF (3 ml) were sucessively added Yb(OTf)$_3$ (0.025 g, 0.04 mmol, 10 mol %) and an enol silyl ether (0.4 mmol) in THF (1 ml) at room temperature. The mixture was stirred for 24 h at this temperature and then the THF was removed under reduced pressure. Water was added and the product extracted with CH$_2$Cl$_2$. After the usual work-up, the crude product was chromatographed on silica gel to yield the aldol product.

Aldol-type reaction between aldehydes and vinyl ethers is promoted by traces of Yb(fod)$_3$ [11]. The primary products, homoallylic alcohols, undergo *in situ* reaction with excess vinyl ethers to provide acetal derivatives.

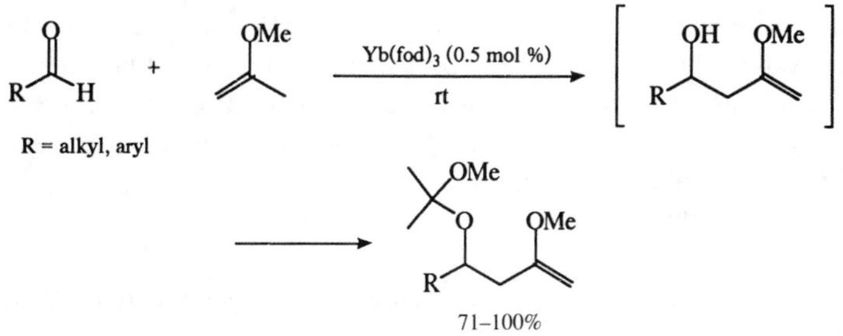

71–100%

Lanthanide alkoxides such as La$_3$(OBu^t)$_9$ serve as efficient catalysts for the aldol reactions between aldehydes and nitro alkanes or α-chloro ketones [12]. It should be noted that remarkably high asymmetric inductions in the nitroaldol reactions have been observed by the use of the asymmetric catalyst prepared from LaCl$_3$·7H$_2$O, dilithium (R)-(+)-binaphthoxide (1 mol equiv.), NaOBu^t(1 mol equiv.) and H$_2$O (4 mol equiv.) [12–14]. An example is shown below [14].

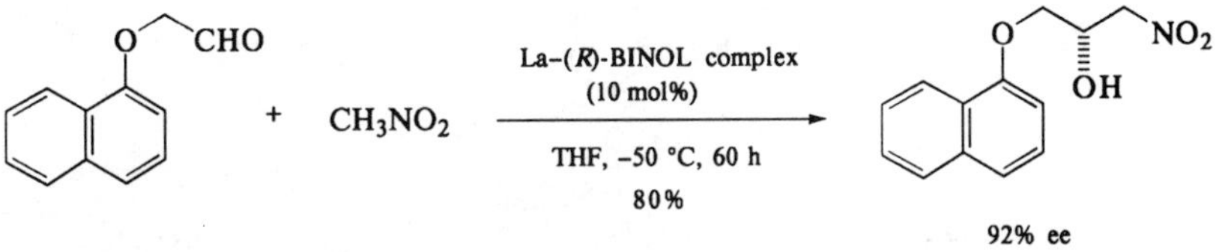

La-(R)-BINOL complex
(10 mol%)

THF, –50 °C, 60 h
80%

92% ee

REFERENCES

1. K. Takai and C. H. Heathcock, *J. Org. Chem.* **50**, 3247 (1985).
2. A. E. Vougioukas and H. B. Kagan, *Tetrahedron Lett.* **28**, 5513 (1987).
3. L. Gong and A. Streitwieser, *J. Org. Chem.* **55**, 6235 (1990).
4. K. Mikami, M. Terada and T. Nakai, *J. Org. Chem.* **56**, 5456 (1991).
5. K. Mikami, M. Terada and T. Nakai, *Tetrahedron Asymm.* **2**, 993 (1991).
6. J.-H. Gu, M. Terada, K. Mikami and T. Nakai, *Tetrahedron Lett.* **33**, 1465 (1992).
7. M. Terada, J.-H. Gu, D.C. Deka, K. Mikami and T. Nakai, *Chem. Lett.* 29 (1992).
8. S. Kobayashi, *Chem. Lett.* 2187 (1991).

9.   S. Kobayashi and I. Hachiya, *Tetrahedron Lett.* **33,** 1625 (1992).
10. S. Kobayashi, I. Hachiya and T. Takahori, *Synlett* 371 (1993).
11. M. V. Deaton and M. A. Ciufolini, *Tetrahedron Lett.* **34,** 2409 (1993).
12. H. Sasai, T. Suzuki, S. Arai, T. Arai, and M. Shibasaki, *J. Am. Chem. Soc.* **114,** 4418 (1992).
13. H. Sasai, T. Suzuki, N. Itoh and M. Shibasaki, *Tetrahedron Lett.* **34,** 851 (1993).
14. H. Sasai, N. Itoh, T. Suzuki and M. Shibasaki, *Tetrahedron Lett.* **34,** 855 (1993).

5.4.3   Cycloaddition reactions

5.4.3.1   Diels–Alder reactions

Lanthanide complexes such as NMR shift reagents effectively catalyse Diels–Alder reactions. Although examples so far reported are limited, use of lanthanide catalysts provides a promising protocol for the Diels–Alder reactions which are difficult to achieve or impossible using other strong Lewis acids. Several examples including intramolecular reactions are tabulated in Table 5.6.

Characteristic features of this method are as follows:

(1) Reactions are catalysed even when there are no oxygen functional groups in the molecules.
(2) The reactions occur under mild conditions, and therefore acid-sensitive functional groups remain unchanged.
(3) Good to high regio- and stereoselectivities are observed, *endo* adducts being major products in most cases.

Eu(fod)$_3$-catalysed addition of furan to ethyl buta-2,3-dienoate [3]

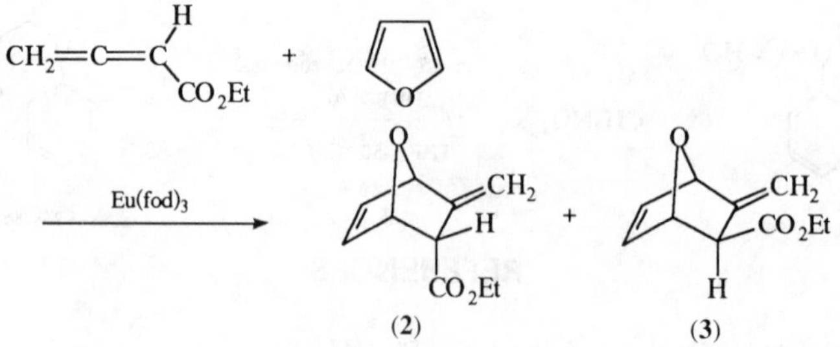

To ethyl buta-2,3-dienoate (0.200 g, 1.78 mmol) was added freshly distilled furan (10 ml) and Eu(fod)$_3$ (0.019 g, 0.018 mmol). The reaction mixture was kept at room temperature for 12 h, after which time excess furan was allowed to evaporate. The residue was chromatographed on silica gel (benzene) to afford compounds **2** (0.205 g, 64%) and **3** (0.052 g, 16%).

REFERENCES

1. T. C. Morrill, R. A. Clark, D. Bilobran and D. S. Youngs, *Tetrahedron Lett.* 397 (1975).
2. S. Danishefsky and M. Bednarski, *Tetrahedron Lett.* **26**, 2507 (1985).
3. M. P. S. Ishar, A. Wali and R. P. Gandhi, *J. Chem. Soc. Perkin Trans.* **1**, 2185 (1990).
4. L. Minuti, L. Radics, A. Taticchi, L. Venturini and E. Wenkert, *J. Org. Chem.*, **55**, 4261 (1990).
5. J. I. Levin, *Tetrahedron Lett.* **30**, 2355 (1989).
6. K. Takeda, Y. Igarashi, K. Okazaki, E. Yoshii and K. Yamaguchi, *J. Org. Chem.* **55**, 3431 (1990).
7. K. Takeda, E. Kawanishi, H. Nakamura and E. Yoshii, *Tetrahedron Lett.* **32**, 4925 (1991).

5.4.3.2 Hetero Diels–Alder reactions

Cycloaddition reaction of dienes with carbonyl compounds represents one of the most significant lanthanide(III)-catalysed processes. The reaction provides highly functionalized dihydropyran derivatives which are versatile as synthetic intermediates. As for dienes, electron-rich olefins such as 1-methoxy-3-(trimethylsiloxy)-1,3-butadiene (Danishefsky's diene) and 1,3-dimethoxy-1-(trimethylsiloxy)-1,3-butadiene (Brassaed's diene) are preferably employed. These dienes are extremely acid labile, and hence strong Lewis acids such as $AlCl_3$ or $TiCl_4$ are not suitable for these reactions. Mild Lewis acidic lanthanide complexes are advantageously employed as the catalysts, since they efficiently activate carbonyl components and enhance the addition reactions, but they do not react with dienes or products such as dihydropyrans which possess highly acid-sensitive functional groups. As the catalysts, NMR shift reagents such as $Eu(hfc)_3$, $Eu(fod)_3$ and $Yb(fod)_3$ are frequently employed, since they are soluble in non-oxygen-containing solvents such as chloroform, toluene and *o*-dichlorobenzene.

The reactions can be conveniently carried out by dissolving the diene and aldehyde in a suitable solvent followed by addition of a lanthanide complex (1–5%). In some cases no solvent is employed and high pressure is suitable for less reactive substrates [1].

As the carbonyl dienophiles, aldehydes are most frequently employed and ketones are rarely used. The method is applicable even to α,β-unsaturated carbonyl compounds [2–4]. The carbon–carbon unsaturated bonds of these substrates are good dienophile components, and Diels–Alder reactions occur in the absence of catalysts. Lanthanide catalysts preferentially promote hetero Diels–Alder reactions rather than Diels–Alder reactions, as exemplified in the following scheme [3].

Most hetero Diels–Alder reactions occur with chemo- and stereoselectivity. A typical example is depicted below, exhibiting *endo* selectivity of the reaction [5].

TABLE 5.6
Diels–Alder Reactions Catalysed by Lanthanide Complexes

Substrate	Catalyst (mol %)	Conditions	Product (yield %)	Ref.
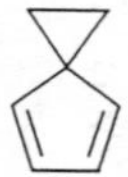 (spiro[2.4]hepta-4,6-diene) + CH$_2$=CHCHO	Eu(tfn)$_3$	0°, 7 days	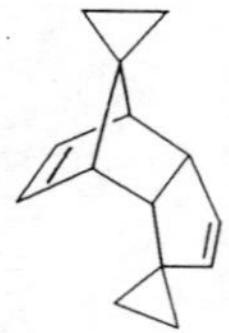 (cycloadduct)	[1]
 (cyclopentadiene) + (furan)	Yb(fod)$_3$ (10)	rt, 24–48 h	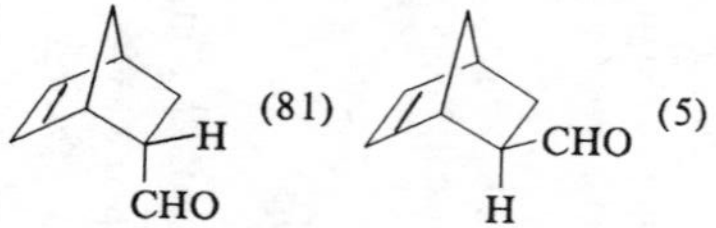 H/CHO (81) and CHO/H (5)	[2]
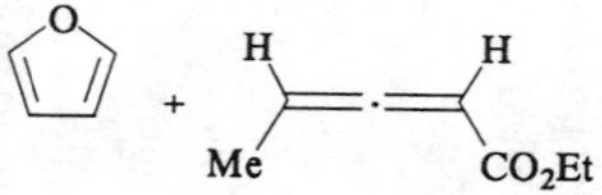 (furan) + H–C=C=C(Me)–CO$_2$Et	Eu(fod)$_3$ (1)	rt, 12 h	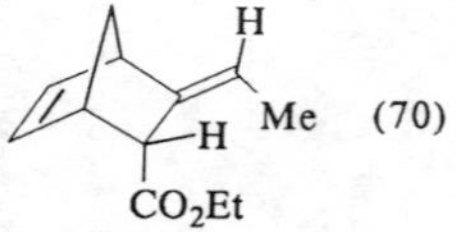 Me, H, CO$_2$Et adduct (70)	[3]
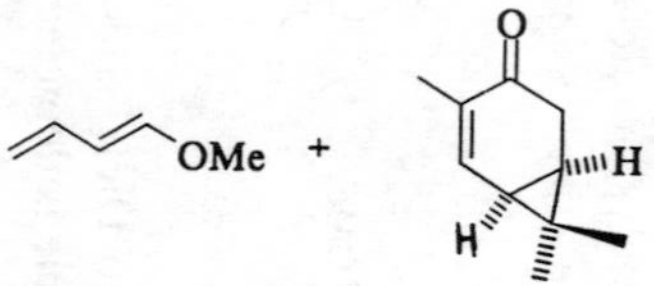 (methylcyclohexenone + gem-dimethyl) + OMe-butadiene	Eu(fod)$_3$ (50)	110 °C, 38 h	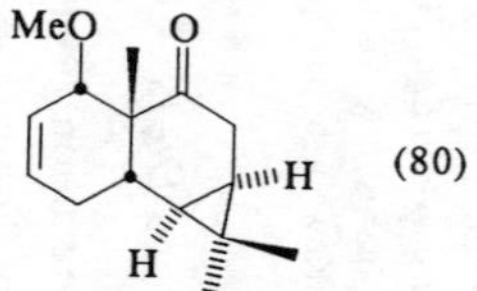 MeO, O fused bicyclic adduct (80)	[4]

Substrate	Catalyst	Conditions	Product (yield %)	Ref.
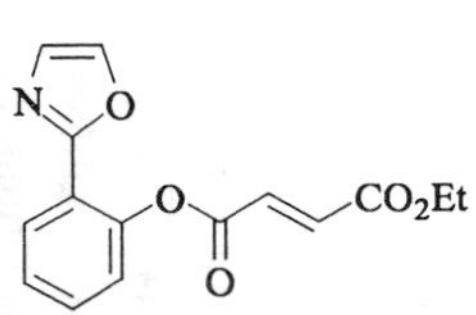 oxazole aryl ester with CO₂Et acrylate side chain	Eu(fod)$_3$ (7)	o-C$_6$H$_4$Cl$_2$, reflux, 16 h	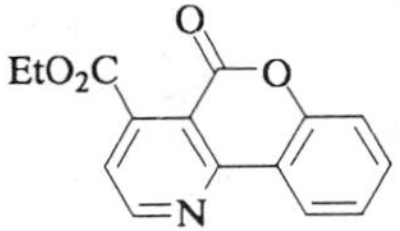 EtO$_2$C-substituted chromeno[?]pyridinone (46)	[5]
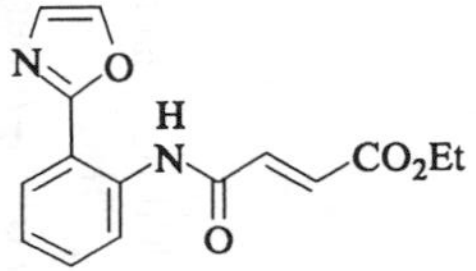 oxazole aryl amide (N–H) with CO₂Et acrylamide side chain	Eu(fod)$_3$ (7)	o-C$_6$H$_4$Cl$_2$, reflux, 16 h	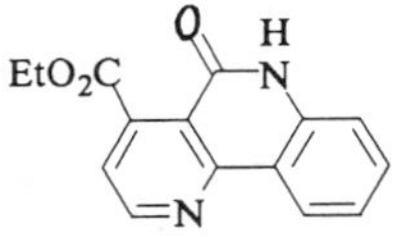 EtO$_2$C-substituted quinolinone (75)	[5]
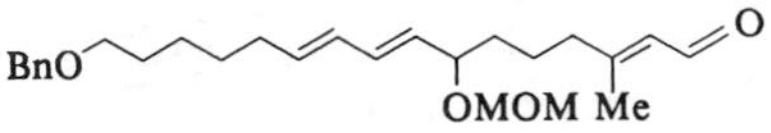 triene aldehyde (BnO, OMOM, Me)	Yb(fod)$_3$ (1)	toluene, reflux	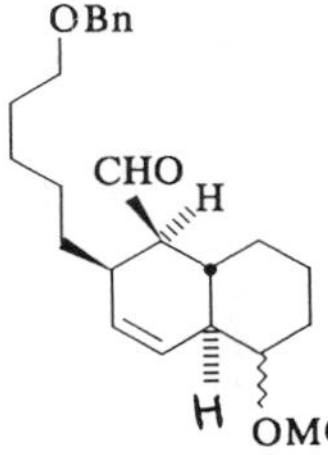 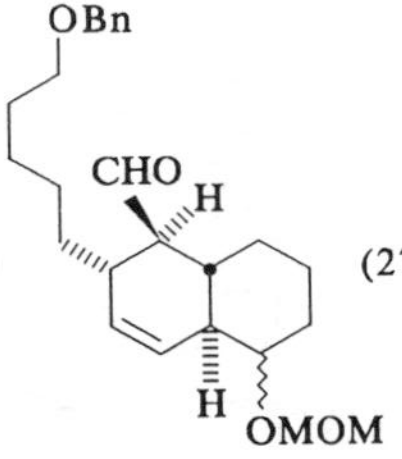 OBn/CHO/OMOM decalin (27) + (60)	[6]
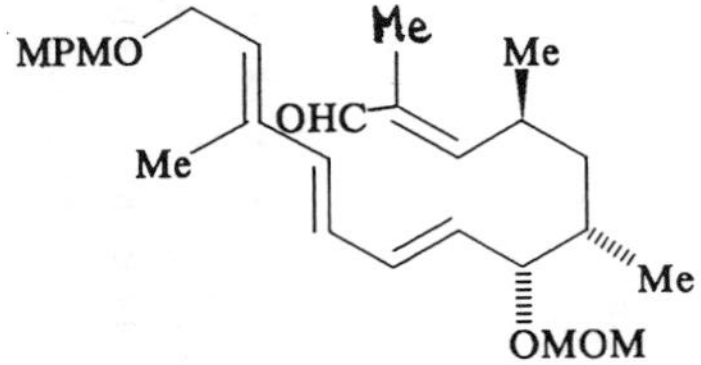 polyene dialdehyde (MPMO, Me, OMOM)	Yb(fod)$_3$ (5)	o-C$_6$H$_4$Cl$_2$, 180 °C, 30 min	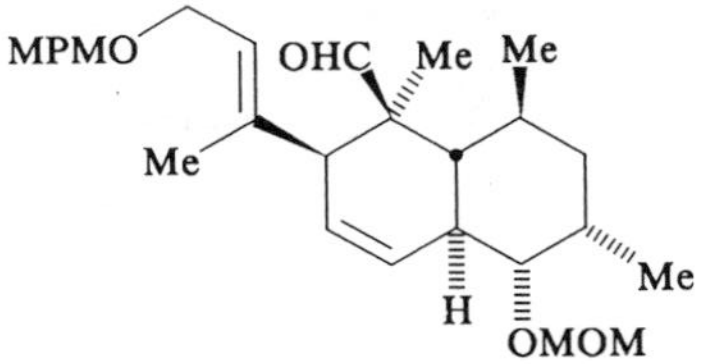 MPMO/OHC/Me/OMOM bicyclic (52)	[7]

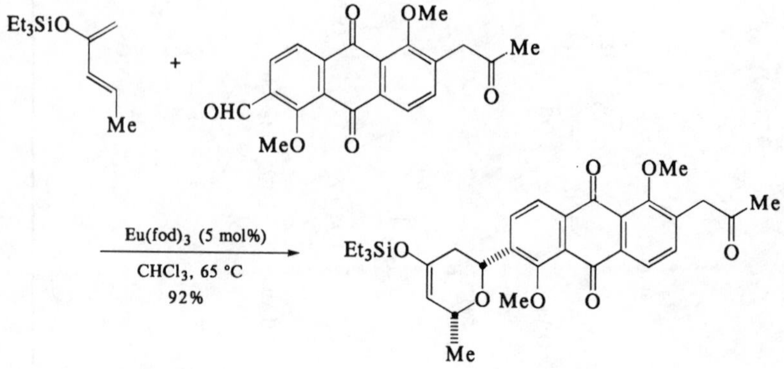

Et$_3$SiO

Me

+

O OMe

Me

OHC

MeO O

O

Eu(fod)$_3$ (5 mol%)

CHCl$_3$, 65 °C

92%

Et$_3$SiO

O OMe

Me

O

MeO O

Me

The *endo* selectivity is quite general and it is observed even in the reaction of aliphatic aldehydes, in that there exists no obvious secondary orbital interaction between the diene and the simple alkyl group [6,7]. This *endo* selectivity is reasonably explained by considering the effective size of the catalyst. The lanthanide(III) cation coordinates with carbonyl oxygen *anti* to the alkyl group of the aldehyde, and the bulky catalyst moiety occupies the *exo* position at the transition state, resulting in the formation of *endo* adduct.

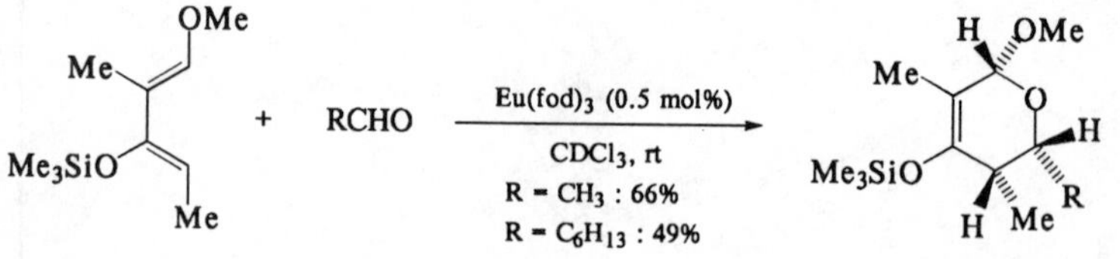

OMe

Me

Me$_3$SiO

Me

+ RCHO

Eu(fod)$_3$ (0.5 mol%)

CDCl$_3$, rt

R = CH$_3$: 66%

R = C$_6$H$_{13}$: 49%

H OMe

Me

O

H

Me$_3$SiO

R

H Me

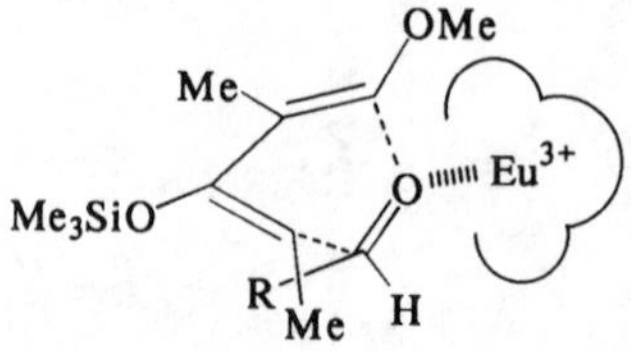

OMe

Me

Me$_3$SiO

O Eu^{3+}

R H

Me

Catalytic amounts of Yb(fod)$_3$ promote the cycloaddition reaction between α,β-unsaturated carbonyl compounds and vinyl ethers [8,9]. It is interesting to note that *endo* selectivity is observed in inverse demand hetero Diels–Alder reactions, as described in the following scheme [8].

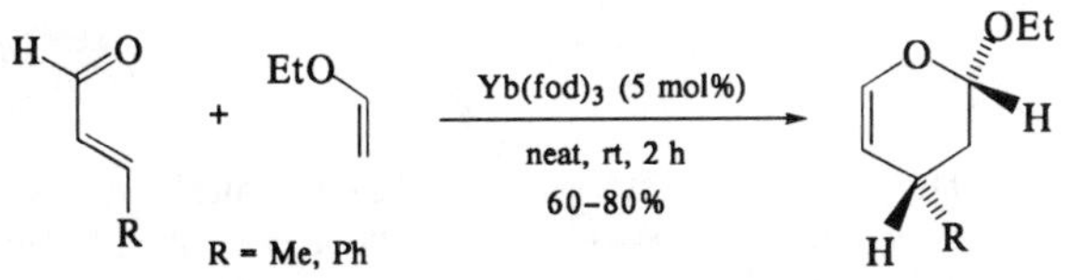

Diastereoselectivity of hetero Diels–Alder reaction has been studied by the research groups of Midland and Danishefsky using α- or β-heterosubstituted aldehydes [10–13]. In general, substrates that are capable of chelation with lanthanide(III) provide chelation controlled products, whereas stereocontrol is determined by Felkin–Anh sense in cases where the substrates are not capable of chelation by steric or electronic effects. Examples of chelation and non-chelation controlled reactions are given below [10].

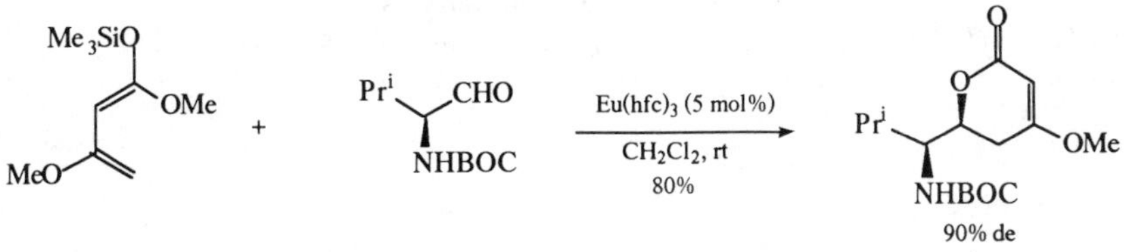

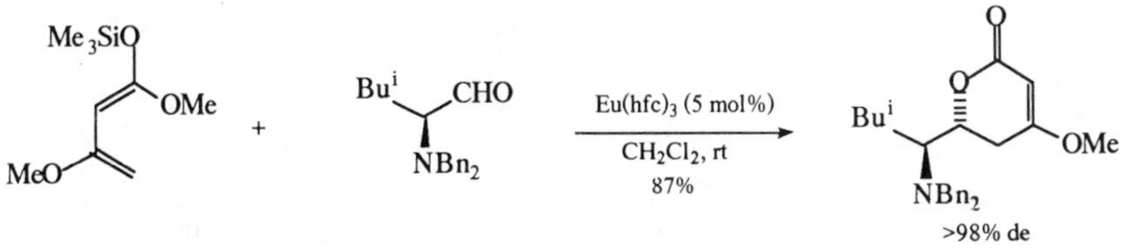

Asymmetric hetero Diels–Alder reactions have been studied by Danishefsky and his coworkers using chiral catalysts and chiral auxiliaries [13,14]. Substantial asymmetric induction up to 58% ee has been observed by the use of (+)-Eu(hfc)$_3$ as a catalyst. The asymmetric induction is affected by substituent patterns on dienes; bulky substituents on positions 1 and 4 tend to give higher enantiomeric excesses. Reactions at lower temperature significantly improve enantioselectivity.

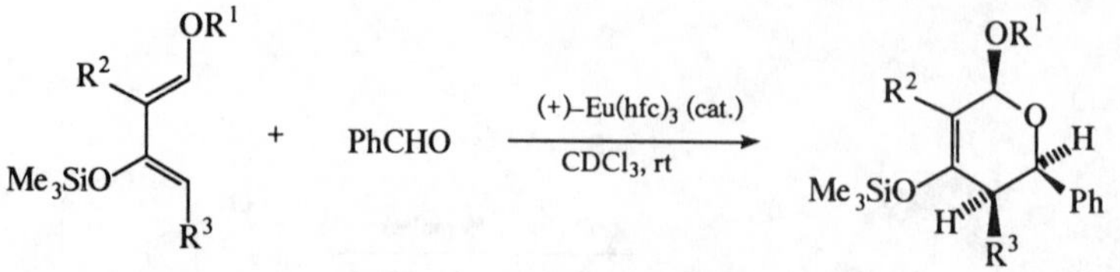

R^1 = Me, R^2 = H, R^3 = H : 18% ee   R^1 = Me, R^2 = Me, R^3 = H : 15% ee
R^1 = Pr^i, R^2 = H, R^3 =H : 28% ee   R^1 = Bu^t, R^2 = Me, R^3 =H : 39% ee
R^1 = Bu^t, R^2 = H, R^3 = H : 38% ee   R^1 = Me, R^2 = Me, R^3 = Me : 36% ee
                                        R^1 = Bu^t, R^2 = Me, R^3 = Me : 42% ee

Modest selectivities are observed in the reactions of chiral dienes with aldehydes in the presence of achiral Eu(fod)$_3$. However, the combination of chiral dienes with chiral (+)-Eu(hfc)$_3$ catalyst exhibits striking interactivities, resulting in some instances in diastereofacial excess of 95% [14]. It is interesting to note that the "mismatched" pair rather than "matched" pair produces a strikingly high diastereomeric ratio. These unpredictable results arise from "specific interactivity" among the dissymmetry elements of the catalyst and the auxiliary.

Synthesis of (2R,6R)-2-(1-phenmenthyloxy)-6-phenyl-4-[(t-butyldimethylsilyl)oxy]-2H(5,6)-dihydropyran (compound 5) [14]

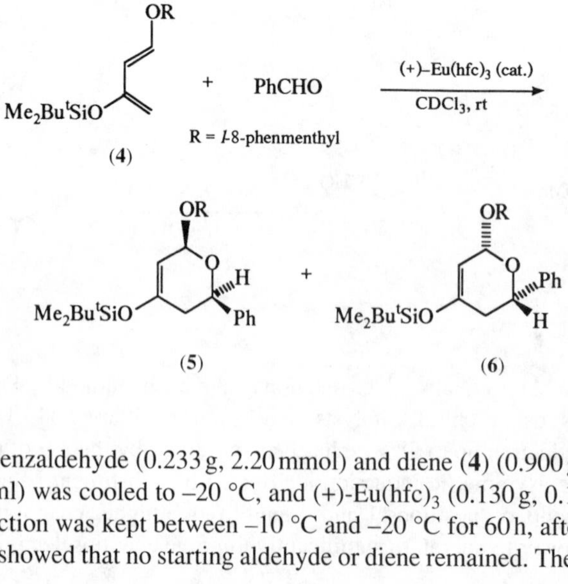

A solution of benzaldehyde (0.233 g, 2.20 mmol) and diene (4) (0.900 g, 2.20 mmol) in hexane (12 ml) was cooled to –20 °C, and (+)-Eu(hfc)$_3$ (0.130 g, 0.11 mmol) was added. The reaction was kept between –10 °C and –20 °C for 60 h, after which time NMR analysis showed that no starting aldehyde or diene remained. The reaction was quenched with triethylamine (8 ml) and methanol (4 ml) and the mixture was allowed to warm to room temperature. The volatiles were removed *in vacuo*, the crude material was passed through a plug of silica gel, and the plug was washed with ethyl acetate. Concentration of the organics *in vacuo* gave enol silyl ethers (5) and (6) in a ratio of 25:1 (1.10 g, 95%). Crystallization of this material from ethanol and recrystallization of the residue from the mother liquors gave optically pure compound (5) as white crystals (0.69 g, 60%); m.p. 59.5–60.8 °C; $[\alpha]_{23D}+47.3°$ (*c* 1.1 in CHCl$_3$).

REFERENCES

1. J. Jurczak, A. Golebiowski and T. Bauer, *Synthesis* 928 (1985).
2. S. Castellino and J. J. Sims, *Tetrahedron Lett.* **25**, 2307 (1984).
3. S. Castellino and J. J. Sims, *Tetrahedron Lett.* **25**, 4059 (1984).
4. M. M. Midland and R. S. Graham, *J. Am. Chem. Soc.* **106**, 4294 (1984).
5. S. J. Danishefsky, B. J. Uang and G. Quallich, *J. Am. Chem. Soc.* **106**, 2453 (1984).
6. M. Bednarski and S. J. Danishefsky, *J. Am. Chem. Soc.* **105**, 3716 (1983).
7. S. J. Danishefsky, W. H. Pearson and D. F. Harvey, *J. Am. Chem. Soc.* **106**, 2456 (1984).
8. S. J. Danishefsky and M. Bednarski, *Tetrahedron Lett.* **25**, 721 (1984).
9. C. Spino and G. Liu, *J. Org. Chem.* **58**, 817 (1993).
10. M. M. Midland and M. M. Afonso, *J. Am. Chem. Soc.* **111**, 4368 (1989).
11. M. M. Midland and R. W. Koops, *J. Org. Chem.* **55**, 5058 (1990).
12. M. M. Midland and R. W. Koops, *J. Org. Chem.* **55**, 4647 (1990).
13. M. Bednarski, C. Maring and S. J. Danishefsky, *Tetrahedron Lett.* **24**, 3451 (1983).
14. M. Bednarski and S. J. Danishefsky, *J. Am. Chem. Soc.* **108**, 7060 (1986).

5.4.3.3 Other cycloaddition reactions

In the presence of lanthanide shift reagents (Yt(fod)$_3$ or Eu(fod)$_3$), ketene imines react with aldehydes in [2+2] cycloaddition manner to provide 2-iminooxetanes as a mixture of *cis* and *trans* stereoisomers [1]. Use of other Lewis acids as catalysts results in oligomerization of ketene imines, and the cycloaddition has been achieved by the use of lanthanide catalysts.

α,α-Dibromo ketones react with cyclic 1,3-dienes in the presence of $SnCl_2$–$CeCl_3$ to afford [3+4] cycloadduct in good to high yields [2]. Under similar conditions, enamines undergo formal [3+2] cycloaddition to give 2-cyclopenten-1-ones after treatment with ethanolic NaOH solution.

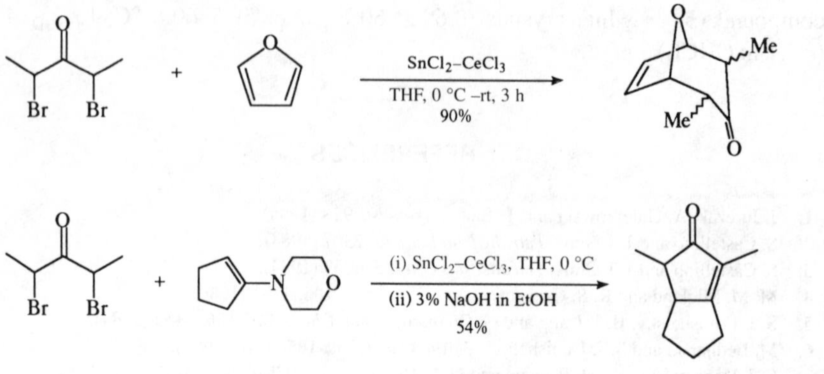

REFERENCES

1. G. Barbaro, A. Battaglia and P. Giorgianni, *J. Org. Chem.* **53,** 5501 (1988).
2. S. Fukuzawa, M. Fukushima, T. Fujinami and S. Sakai, *Bull. Chem. Soc. Jpn* **62,** 2348 (1989).

5.4.4 Hydrocyanation of carbonyl compounds and related reactions

Lanthanide(III) alkoxides such as $La(OPr^i)_3$, $Ce(OPr^i)_3$, $Sm(OPr^i)_3$ and $Yb(OPr^i)_3$ are efficient catalysts for the transhydrocyanation of acetone cyanohydrin to aldehydes or ketones [1]. This method is useful as a practical cyanating method in the laboratory, since it avoids the use of highly toxic hydrogen cyanide.

$$R^1COR^2 + Me_2C(OH)CN \xrightarrow{Ln(OPr^i)_3\ (catalyst)} R^1\!-\!C(R^2)(OH)\!-\!CN + Me_2CO$$

Preparation of mandelonitrile from benzaldehyde and acetone cyanohydrin [1]

To a mixture of benzaldehyde (0.265 g, 2.5 mmol) and acetone cyanohydrin (0.255 g, 3 mmol) was added $Yb(OPr^i)_3$ (0.025 mmol, 0.083 ml of 0.3 M THF solution) under argon, and the mixture was stirred for 3 min at room temperature. The reaction was quenched with 2 M HCl, and the organic materials extracted with ether. The combined organic layers were dried over Na_2SO_4, and concentrated *in vacuo*. The crude product was purified by chromatography on silica gel (hexane–ethyl acetate, 10 : 1) to afford mandelonitrile (0.27 g, 80%).

Ytterbium(III) cyanide promotes the addition reaction of cyanotrimethylsilane to carbonyl compounds [2]. The method is applicable to a variety of carbonyl compounds including enolizable ketones and α,β-unsaturated ketones. Reactions with substituted cyclohexanones procced in a highly stereoselective manner.

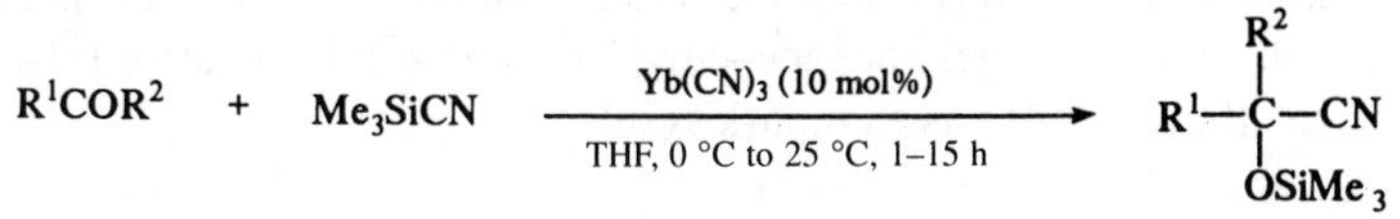

$$R^1COR^2 \quad + \quad Me_3SiCN \quad \xrightarrow[\text{THF, 0 °C to 25 °C, 1–15 h}]{\text{Yb(CN)}_3 \text{ (10 mol\%)}} \quad R^1-\underset{\underset{OSiMe_3}{|}}{\overset{\overset{R^2}{|}}{C}}-CN$$

Reaction of acetophenone with cyanotrimethylsilane in the presence of Yb(CN)₃ [2]

To a mixture of Yb(CN)$_3$ (0.050 mg, 0.2 mmol) and cyanotrimethylsilane (0.238 g, 2.4 mmol) in THF (5 ml), acetophenone (0.240 g, 2 mmol) was added at room temprature and the mixture stirred for 1 h. Hydrolytic work-up gave 2-phenyl-2-trimethylsiloxypropionitrile (0.434 g, 99%).

REFERENCES

1.   H. Ohno, A. Mori and S. Inoue, *Chem. Lett.* 375 (1993).
2.   S. Matsubara, T. Takai and K. Utimoto, *Chem. Lett.* 1447 (1991).

5.4.5   Ring opening reactions

Ytterbium(III) cyanide and lanthanide(III) alkoxides serve for ring cleavage of oxiranes and aziridines with cyanotrimethylsilane [1–3]. The reactions take place with high regio- and stereoselectivities to give β-trimethylsiloxy nitriles and β-aminonitrile derivatives.

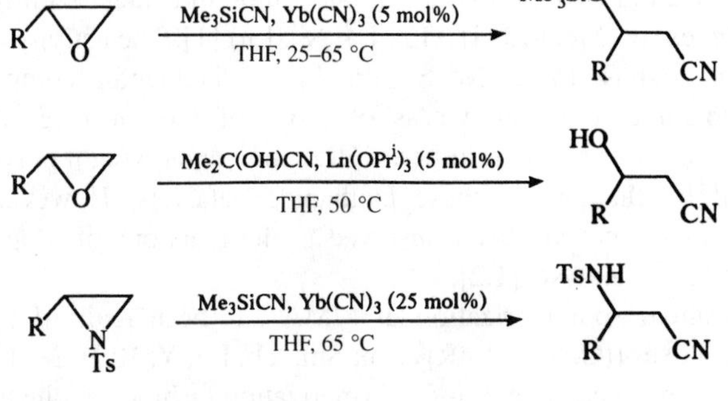

Reaction of 1,2-epoxyoctane with acetone cyanohydrin in the presence of La(OPr^i)$_3$ [2]

To a mixture of 1,2-epoxyoctane (0.32 g, 2.5 mmol) and acetone cyanohydrin (0.26 g, 3.0 mmol) was added La(OPr^i)$_3$ (0.125 mmol, 0.42 ml of 0.3 M THF solution) under argon, and the mixture was stirred for 1 h at 50 °C. The reaction was quenched with 10 ml of water, and the organic materials were extracted twice with 50 ml of ether. The combined organic layers were dried over anhydrous Na$_2$SO$_4$ and concentrated *in vacuo* to leave a crude product, which was purified by chromatography on silica gel (hexane–ethyl acetate, 4 : 1) to afford 1-cyano-2-hydroxyoctane (0.31 g, 80%) as a colourless oil.

Reaction of N-tosylcyclohexeneimine with cyanotrimethylsilane in the presence of Yb(CN)$_3$ [3]

A mixture of *N*-tosylcyclohexeneimine (0.535 g, 2 mmol), cyanotrimethylsilane (0.40 g, 4 mmol) and Yb(CN)$_3$ (0.13 g, 0.5 mmol) in THF (7 ml) was stirred at 65 °C for 2.5 h. The reaction mixture was treated with water, and extracted with ether. The extract was washed with brine, dried over Na$_2$SO$_4$ and concentrated. The residue was subjected to column chromatography on silica gel (hexane–ethyl acetate) to afford *trans*-2-tosylaminocyclohexanecarbonitrile (0.501 g, 90%).

REFERENCES

1. S. Matsubara, H. Onishi and K. Utimoto, *Tetrahedron Lett.* **31,** 6209 (1990).
2. H. Ohno, A. Mori and S. Inoue, *Chem. Lett.* 975 (1993).
3. S. Matsubara, T. Kodama and K. Utimoto, *Tetrahedron Lett.* **31,** 6379 (1990).

5.4.6 Polymerizations

Organolanthanide(III) complexes possessing pentamethylcyclopentadienyl groups are extremely active homogeneous catalysts for the polymerization of ethylene, methyl methacrylate (MMA), ε-caprolactone and δ-valerolactone. Ethylene is rapidly polymerized in living polymerization fashion in contact with lanthanide(III) hydride complexes [(C$_5$Me$_5$)$_2$LnH]$_2$ (Ln: La, Nd, Lu) [1]. The activity order follows decreasing ionic radius: La ≥ Nd >> Lu. Earlier lanthandie complexes exhibit activities comparable to or in excess of those of the most reactive ethylene polymerization catalysts reported hitherto. High molecular weight polyethylene can be synthesized by the use of these lanthanide catalysts. However, significant polymerizations have not yet been observed in the reactions of propylene or 1,3-butadiene with these catalysts [1,2].

The most dramatic polymerization of MMA has been realized by the use of lanthanide complexes [(C$_5$Me$_5$)$_2$LnR]$_2$ (Ln: Sm, Yb, Lu, Y; R: H, Me) [3,4]. The polymerization proceeds in a living polymerization fashion, as illustrated below.

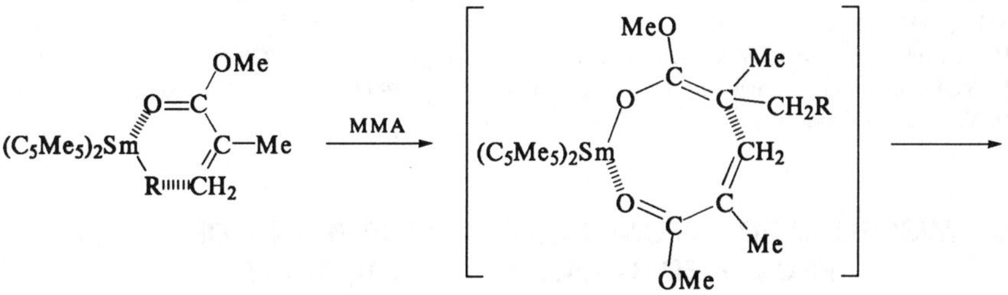

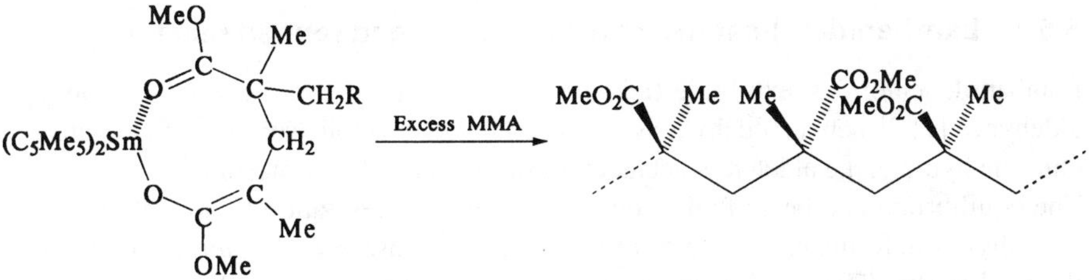

The following are the characteristic features of this lanthanide-catalysed polymerization of MMA.

(1) Monodispersed polymers of high molecular weight (exceeding 5×10^5) are obtained.
(2) The resultant polydispersities, M_w/M_n, are extremely narrow and typically are 1.02.
(3) The efficiency of the catalyst reaches 95% and polymerizations occur rapidly.
(4) High syndiotacticity (> 95%) is realized by the reaction at temperature below $-95\,°C$.

Polymerization of MMA catalysed by $[(C_5Me_5)_2SmH]_2$ [5]

Under argon atmosphere, completely dried MMA (4 ml) and dry toluene (40 ml) were placed in a 100 ml Schlenk tube containing a stirrer bar. The tube was immersed in an ice bath, and a solution of $[(C_5Me_5)_2SmH]_2$ (0.042 g, 0.1 mmol) in toluene (1 ml) was added all at once with vigorous stirring. After being stirred for 2 h, the reaction mixture was poured into ethanol (500 ml) and the mixture left to stand for 2–3 h. The precipitated solid was collected by filtration and dried to give poly(MMA) (4.9 g, 99%), M_n 52 000, M_w/M_n 1.03.

REFERENCES

1. G. Jeske, H. Lauke, H. Mauermann, P. N. Swepston, H. Schumann and T. J. Marks, *J. Am. Chem. Soc.* **107**, 8091 (1985).
2. P. L. Watson and G. W. Parshall, *Acc. Chem. Res.* **18**, 51 (1985).
3. H. Yasuda, H. Yamamoto, K. Yokota, S. Miyake and A. Nakamura, *J. Am. Chem. Soc.* **114**, 4908 (1992).

4. H. Yasuda, M. Furo, H. Yamamoto, A. Nakamura, S. Miyake and N. Kibino, *Macromolecules* **25,** 5115 (1992).
5. H. Yasuda, in *Practical Organometallic Chemistry for Synthetic Chemists,* ed. F. Sato, K. Yamamoto and T. Imamoto, p. 244. Kodansha, Tokyo, 1992 (reproduced by permission from H. Yasuda and Kodansha).

5.5 MISCELLANEOUS FUNCTIONAL GROUP TRANSFORMATIONS PROMOTED BY LANTHANIDE(III) SALTS

5.5.1 Lanthanide(III)-catalysed acetalization and related reactions

Lanthanide chlorides exhibit efficient catalytic activity for the acetalization of aldehydes [1]. When an aldehyde is dissolved in a methanolic solution of lanthanide chloride hydrate, the aldehyde–acetal equilibrium is reached almost instantaneously. The equilibrium can be shifted to optimum acetal concentration by the addition of trimethyl orthoformate. The yields of the acetals increase with the atomic number of the lanthanides. Thus, lighter lanthanide chlorides such as $LaCl_3$, $CeCl_3$ and $NdCl_3$ are effective for the synthesis of the acetals of aliphatic aldehydes, and heavy lanthanide chlorides such as $YbCl_3$ are most effective for aromatic aldehydes.

$$R\text{-}CHO \quad + \quad MeOH \quad \xrightarrow[HC(OMe)_3]{LnCl_3} \quad R\text{-}CH(OMe)_2$$

The reactions are carried out under mild conditions and are especially valuable for acid-sensitive substances. Representative examples including chemoselective acetalization are shown below [2–4].

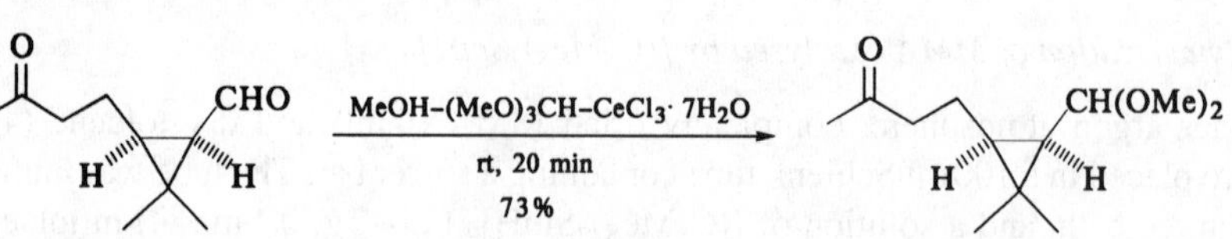

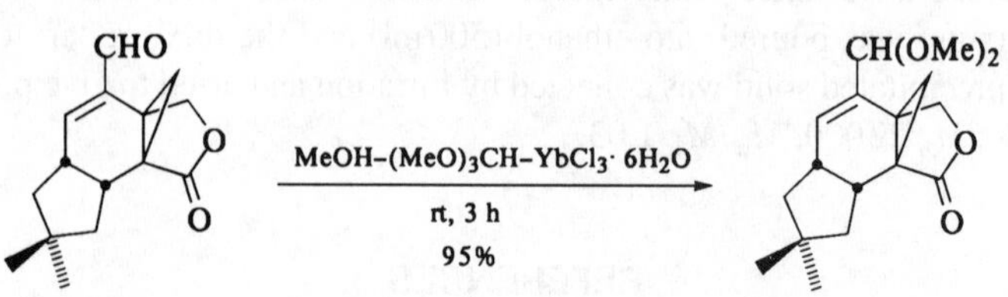

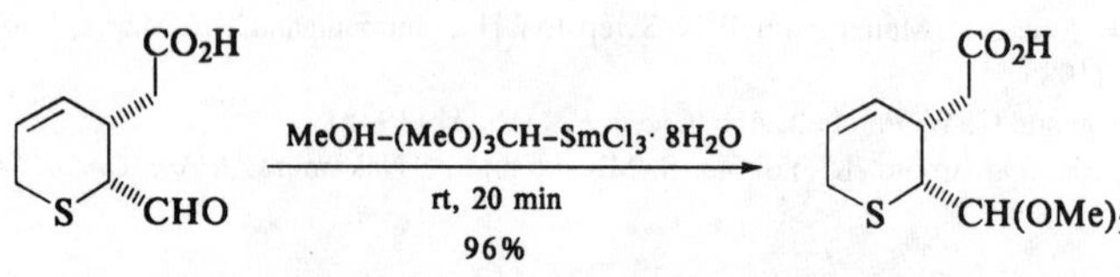

Anhydrous lanthanum chloride promotes thioacetalization of carbonyl compounds [5]. Aldehydes and simple aliphatic ketones are smoothly thioacetalized by 1,2-ethanedithiol at room temperature. However, aromatic ketones and sterically hindered ketones are resistant to thioacetalization under these conditions.

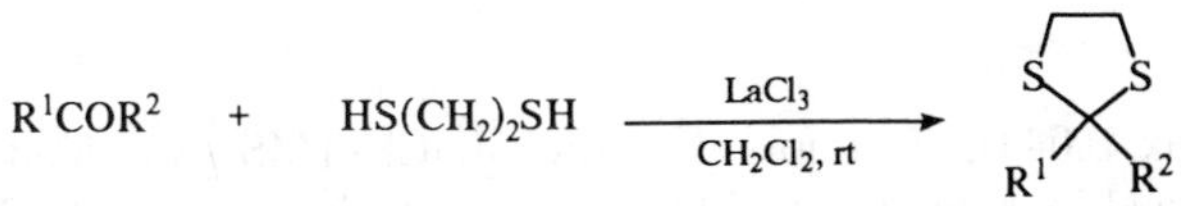

Lanthanide(III) chloride in conjunction with chlorotrimethylsilane serves also as an effective catalyst for selective cleavage of ketone acetals [6]. Under these conditions, aldehyde acetals, t-butyldimethylsilyl ethers, methoxymethyl ethers, benzyl ethers and methyl esters remain unchanged, and therefore this protocol provides a useful entry for chemoselective cleavage of ketone acetals.

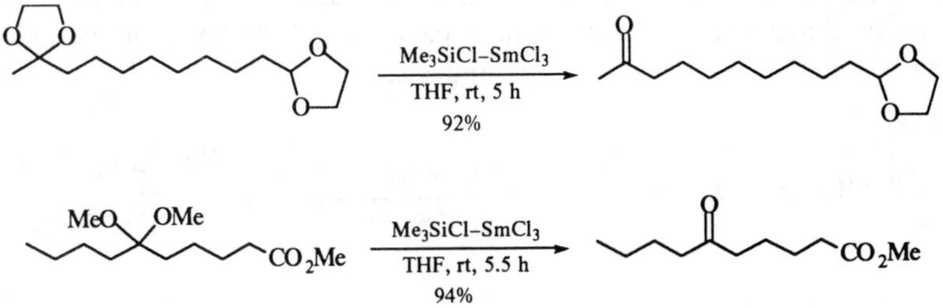

REFERENCES

1. J.-L. Luche and A. L. Gemal, *J. Chem. Soc. Chem. Commun.* 976 (1978).
2. M. D. Taylor, G. Minaskanian, K. N. Winzenberg, P. Santone and A. B. Smith III, *J. Org. Chem.* **47,** 3960 (1982).
3. Y. Tobe, D. Yamashita, T. Takahashi, M. Inata, J. Sato, K. Kakiuchi, K. Kobiro and Y. Odaira, *J. Am. Chem. Soc.* **112,** 775 (1990).
4. K. R. Lawson, B. P. McDonald, O. S. Mills, R. W. Steele, J. K. Sutherland, T. J. Wear, A. Brewster and P. R. Marsham, *J. Chem. Soc. Perkin Trans. 1,*663 (1988).
5. L. Garlaschelli and G. Vidari, *Tetrahedron Lett.* **31,** 5815 (1990).
6. Y. Ukaji, N. Koumoto and T. Fujisawa, *Chem. Lett.* 1623 (1989).

5.5.2 Cyclization reactions

Marks and his coworkers found that organolanthanide complexes of the type $(C_5Me_5)_2LnR$ (R: H,CH(SiMe_3)_2; Ln: La, Nd, Sm, Lu) are highly reactive with respect to cyclization of amino olefins. By this method, variously substituted 5-, 6- and 7-membered secondary amines can be synthesized from amino olefins.

Synthesis of 2-methylpyrrolidine [1]

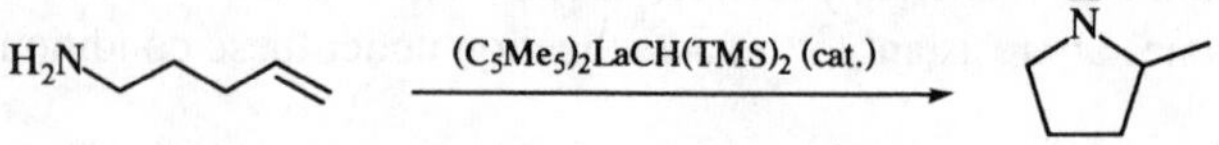

In a glove box, 0.043 g (76 µmol) of $(C_5Me_5)_2LaCH(TMS)_2$ was loaded into a 15 ml round-bottomed flask equipped with a magnetic stir bar. At −78 °C, 2 ml of pentane was vacuum transferred onto the catalyst followed by 2.2 ml of 5-amino-1-pentene (1.69 g, 19.9 mmol). The clear, colourless solution was stirred under argon for 24 h at ambient temperature. The reaction mixture was then freeze-thaw degassed and the volatiles vacuum transferred into a separate flask. Pentane was removed on the rotary evaporator at 0 °C to give 2-methylpyrrolidine (1.45 g, 86%, >95% pure by GC/MS).

Catalytic amounts of lanthanide(III) triflates or perchlorates promote reactions between amines and nitrile to give condensation products [2,3]. Secondary alicyclic amines or dimethylamine react with excess acetonitrile to yield pyrimidines; amidines and triazines being major by-products.

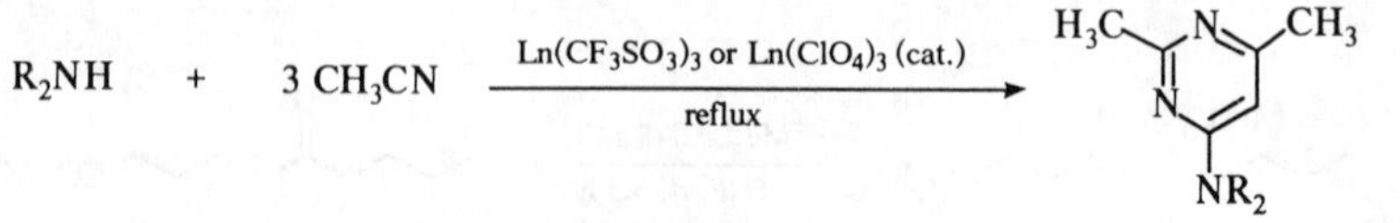

REFERENCES

1. M. R. Gangé, C. L. Stern and T. J. Marks, *J. Am. Chem. Soc.* **114**, 275 (1992).
2. J. H. Forsberg, T. M. Balasubramanian and V. T. Spaziano, *J. Chem. Soc. Chem. Commun.* 1060 (1976).
3. J. H. Forsberg, V. T. Spaziano, T. M. Balasubramanian, G. K. Liu, S. A. Kinsley, C. A. Duckworth, J. J. Poteruca, P. S. Brown and J. L. Miller, *J. Org. Chem.* **52**, 1017 (1987).

5.5.3 Rearrangements

An oxaspiropentane derivative rearranges to a cyclobutanone derivative on treatment with a catalytic amount of $Eu(fod)_3$ in $CDCl_3$ [1]. This ring expansion reaction occurs also using HBF_4 or $LiClO_4$, although $Eu(fod)_3$ gives the highest stereoselectivity.

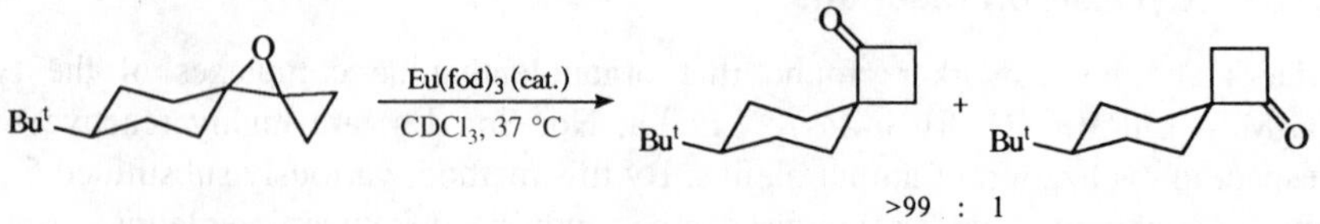

>99 : 1

Epoxides are converted to ketones by treatment with catalytic amounts of Bu^tOSmI_2 (2). Scope of the reaction is limited to monosubstituted epoxides; 1,2- or 2,2-disubstituted epoxides do not undergo the analogous transformation.

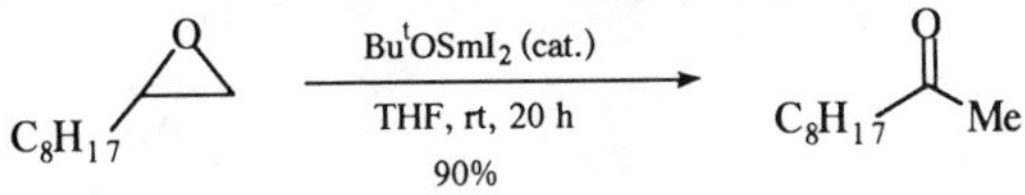

An α-hydroxy ketone rearranges with ring expansion by samarium(III) species, as shown in the following scheme [3]. The reaction is stereospecific and another α-hydroxy ketone epimer does not undergo similar transformation.

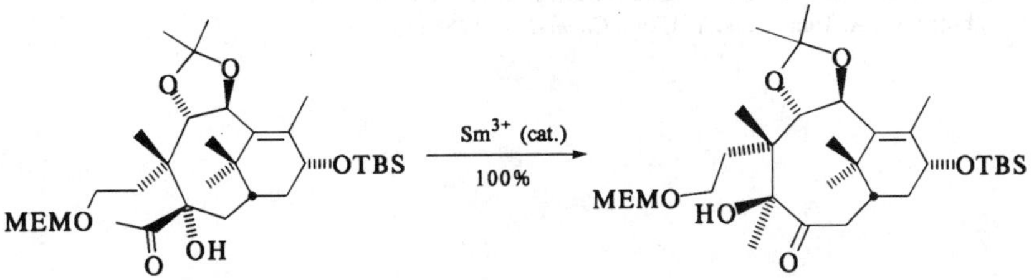

REFERENCES

1. B. M. Trost and M. J. Bogdanowicz, *J. Am. Chem. Soc.* **95**, 5321 (1973).
2. J. Prandi, J. L. Namy, G. Menoret and H. B. Kagan, *J. Organomet. Chem.* **285**, 449 (1989).
3. R. A. Holton and A. D. Williams, *J. Org. Chem.* **53**, 5981 (1988).

5.5.4 Other functional group transformations

Several examples of lanthanide(III)-promoted nucleophilic substitution reactions and nucleophilic additions of iodide ion to activated triple bonds are shown below [1–3].

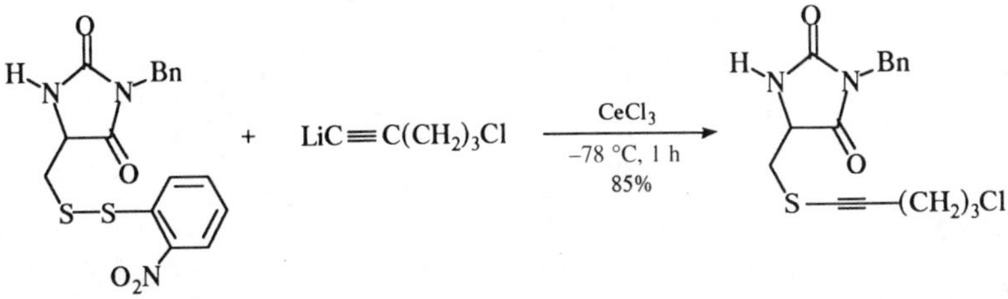

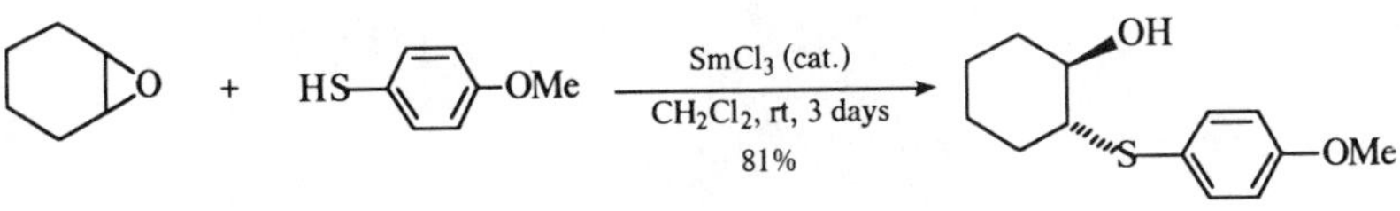

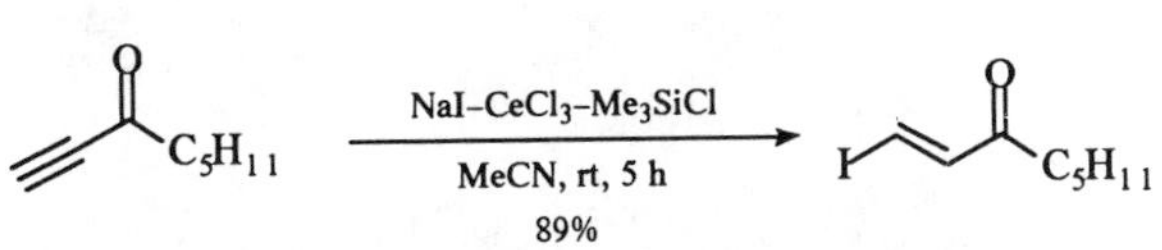

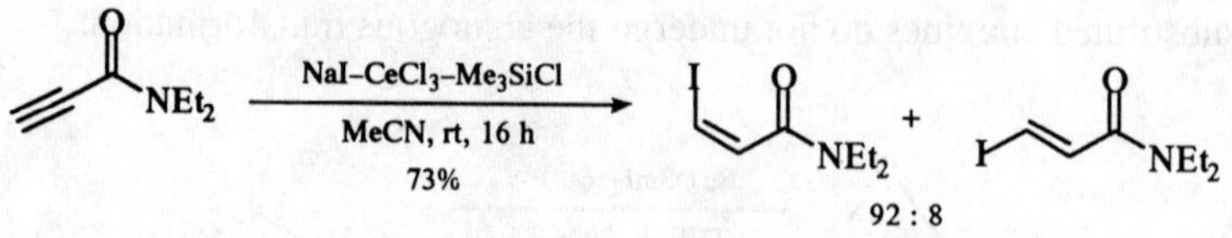

REFERENCES

1. E. J. Corey and M. M. Mehrotra, *Tetrahedron Lett.* **29,** 57 (1988).
2. A. E. Vougioukas and H. B. Kagan, *Tetrahedron Lett.* **28,** 6065 (1987).
3. T. Fujisawa, A. Tanaka and Y. Ukaji, *Chem. Lett.* 1255 (1989).

–6–

Tetravalent Lanthanides

6.1 GENERAL ASPECTS

Cerium, praseodymium, neodymium, terbium and dysprosium can form tetrapositive oxidation states. Cerium(IV), whose electronic configuration is the same as that of xenon, is the most stable, and many cerium(IV) compounds have been isolated. Other tetrapositive lanthanides are extremely unstable, not being capable of existing as isolable salts.

Although all tetrapositive lanthanides have potential one-electron oxidizing power, cerium(IV) salts are almost exclusively employed in synthesis because of their easy availability. Cerium(IV) ammonium nitrate (CAN), $Ce(NH_4)_2(NO_3)_6$, cerium(IV) ammonium sulphate (CAS), $Ce(NH_4)_4(SO_4)_4$, cerium(IV) sulphate, $Ce(SO_4)_2$, cerium(IV) oxide, CeO_2, and cerium(IV) hydroxide, $Ce(OH)_4$, are commercially available at moderate prices, and of these CAN and CAS are most frequently employed in oxidation of organic functional groups [1].

Other lanthanide(IV) compounds have also been reported, for example:

cerium(IV) acetate, $Ce(OCOCH_3)_4$ [2]
cerium(IV) trifluoroacetate, $Ce(OCOCF_3)_4$ [3]
cerium(IV) methanesulphonate, $Ce(OSO_2CH_3)_4$ [4]
cerium trifluoromethanesulphonate, $Ce(OSO_2CF_3)_4$ [5,6]
tetrakis(2,4-pentanedionato)cerium(IV) (cerium(IV) acetylacetonate,
  $Ce(C_5H_7O_2)_4$ [7]
tetrakis(1,1,1-trifluoro-2,4-pentanedionato)cerium(IV) (cerium(IV)
  trifluoroacetylacetonato), $Ce(C_5H_4F_3O_2)_4$ [8]
tris[trinitratocerium(IV)] paraperiodate, $[Ce(NO_3)_3]_3H_2IO_6$ [9]
bis[trinitratocerium(IV)] chromate, $[Ce(NO_3)_3]_2CrO_4$ [10,11]
dinitratocerium(IV) chromate, $Ce(NO_3)_2CrO_4$ [12]
trihydrocerium(IV) hydroperoxide, $Ce(OH)_3OOH$ [13]
dipyridinium hexachlorocerate, $(C_5H_6N)_2CeCl_6$ [14]
bis(triethylammonium) hexakis(nitrato)cerate, $(Et_3NH)_2Ce(NO_3)_6$ [15]
tetrabutylammonium cerium(IV) nitrate, $(Bu_4N)_2Ce(NO_3)_6$ [16]
tetrachlorobis(triphenylphosphine oxide)cerium(IV), $CeCl_4(OPPh_3)_2$ [17]

These compounds possess oxidizing ability different from CAN and CAS, and specific oxidations are possible by the use of these compounds.

The oxidizing abilities of lanthanide(IV) compounds are largely dependent on the reaction medium. The oxidation potential usually increases with the increase of the concentration of acid, depending on the nature of the acid. Cerium(IV) is usually subjected to hydrolysis in aqueous medium, and the coordination with anionic species decreases the oxidation potential. Relatively high oxidation potential is observed in perchloric acid or trifluoromethanesulphonic acid compared with sulphuric acid or nitric acid. This is ascribed to the higher coordination ability of sulphate and nitrate to Ce^{4+} ion.

Oxidizing ability is changed by absorbing the Ce(IV) compounds on supporters such as silica gel, alumina and charcoal. Reagent systems combined with other oxidizing agents have been devised. For example, the CAN–$NaBrO_3$ reagent system exhibits a unique reactivity differing from CAN itself. Other oxidations using catalytic amounts of cerium(IV) compounds are also possible by combination with electrochemical methods.

REFERENCES

1. T.-L. Ho, in *Organic Syntheses by Oxidation with Metal Compounds*, ed. W. J. Mijs and C. R. H. L. de Jonge, p. 569. Plenum, New York, 1986.
2. N. Hay and J. K. Kochi, *J. Inorg. Nucl. Chem.* **30**, 884 (1968).
3. R. O. C. Norman, C. B. Thomas and P. J. Ward, *J. Chem. Soc. Perkin Trans. 1* 2914 (1973).
4. R. P. Kreh, R. M. Spotnitz and J. T. Lundquist, *J. Org. Chem.* **54**, 1526 (1989).
5. R. P. Kreh, R. M. Spotnitz and J. T. Lundquist, *Tetrahedron Lett.* **28**, 1067 (1987).
6. T. Imamoto, Y. Koide and S. Hiyama, *Chem. Lett.* 1445 (1990).
7. T. J. Pinnavaia and R. C. Fay, *Inorg. Syn.* **12**, 77 (1970).
8. T. J. Pinnavaia and R. C. Fay, *Inorg. Syn.* **12**, 79 (1970).
9. H. Firouzabadi, N. Iranpoor, G. Hajipoor and J. Toofan, *Synth. Commun.* **14**, 1033 (1984).
10. H. Firouzabadi, N. Iranpoor, H. Parham and J. Toofan, *Synth. Commun.* **14**, 631 (1984).
11. H. Firouzabadi, N. Iranpoor, H. Parham, A. Sardarian and J. Toofan, *Synth. Commun.* **14**, 717 (1984).
12. H. Firouzabadi, N. Iranpoor, F. Kiaeezadeh and J. Toofan, *Synth. Commun.* **14**, 973 (1984).
13. H. Firouzabadi and N. Iranpoor, *Synth. Commun.* **14**, 875 (1984).
14. D. C. Bradley, A. K. Chatterjee and W. Wardlaw, *J. Chem. Soc.* **1956**, 2260.
15. H. Firouzabadi and N. Iranpoor, *Synth. Commun.* **13**, 1143 (1983).
16. H. A. Muathen, *Indian J. Chem.* **30B**, 522 (1991).
17. F. Brezina, *Inorg. Syn.* **23**, 179 (1985).

6.2   OXIDATIONS OF HYDROCARBONS

6.2.1   Aromatic compounds

Polynuclear aromatics are oxidized by CAN [1], CAS [2,3] and $Ce(OSO_2CH_3)_4$ [4]. In general, symmetrical aromatics such as naphthalene and anthracene are converted to quinones in reasonable yields, while asymmetric ones lead to many products.

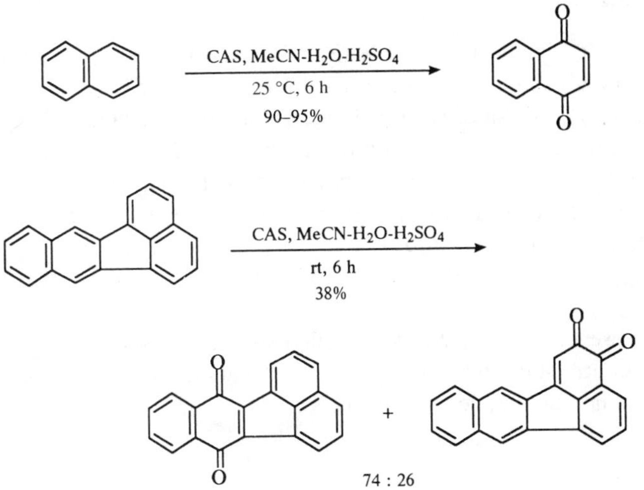

$$\text{CAS, MeCN-H}_2\text{O-H}_2\text{SO}_4$$
$$25\ °C, 6\ h$$
$$90–95\%$$

$$\text{CAS, MeCN-H}_2\text{O-H}_2\text{SO}_4$$
$$\text{rt, 6 h}$$
$$38\%$$

74 : 26

Cerium(IV) ammonium sulphate (CAS) oxidation of polycyclic aromatic hydrocarbons. General procedure [2]

To a solution (or suspension) of the organic substrate (1 mmol) in acetonitrile and 4 normal sulphuric acid (40 ml/10 ml), CAS (6 mmol) in 4 normal sulphuric acid (50 ml) was added and the mixture stirred at 25–50 °C. The cerous salt precipitated as the reaction proceeded. After completion of the reaction, the solution was decanted into a separatory funnel, diluted with water and extracted with ether. The solvent was removed under reduced pressure. The residue was purified in the usual manner to give the product quinones.

Oxidative functionalization of aromatic rings can be performed by the modification of the reaction medium. Nitration of naphthalene derivatives, anthracene and phenanthrene occurs when they are treated with silica gel-supported CAN [5]. Although analogous oxidative nitration occurs in solution, this method provides better yield and high selectivity. Direct alkoxylation of anthracene is performed by treatment with $Ce(OCOCF_3)_4$ in alcohol [6]. A considerable amount of anthraquinone is formed as a major by-product.

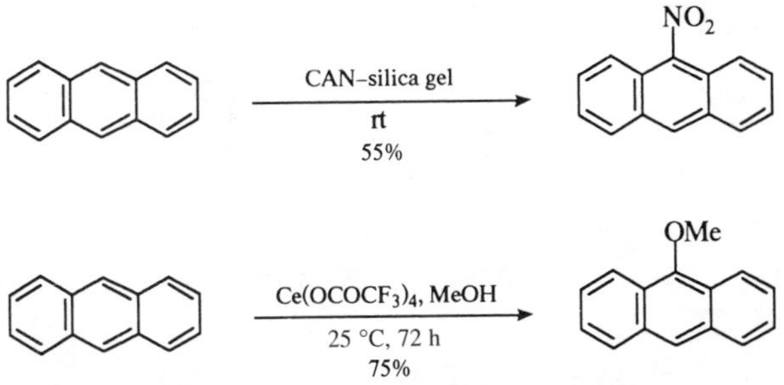

$$\text{CAN–silica gel}$$
$$\text{rt}$$
$$55\%$$

$$Ce(OCOCF_3)_4, \text{MeOH}$$
$$25\ °C, 72\ h$$
$$75\%$$

Reaction of anthracene with silica gel-supported CAN; preparation of
9-nitroanthracene [5]

Anthracene (1 g, 5 mmol) and CAN (3 g, 5.5 mmol) were separately dissolved in
acetonitrile (4 ml) and these solutions each mixed to a slurry with silica gel (2 g and
5 g, respectively). The slurries were dried in a hot-air oven at 60–65 °C for 1 h. The
dried masses were mixed well and this mixture was added to a prepared column of
silica gel (80 g). The column was eluted with petroleum ether–benzene (9 : 1).
Concentration of the eluate afforded 9-nitroanthracene (0.72 g, 55%) in a TLC pure
state as yellow needles, m.p. 146 °C.

In the presence of CAN, polymethylbenzenes, polymethoxybenzene and
naphthalene are subjected to aromatic iodonation with alkali metal iodides,
tetrabutylammonium iodide or iodine [7,8].

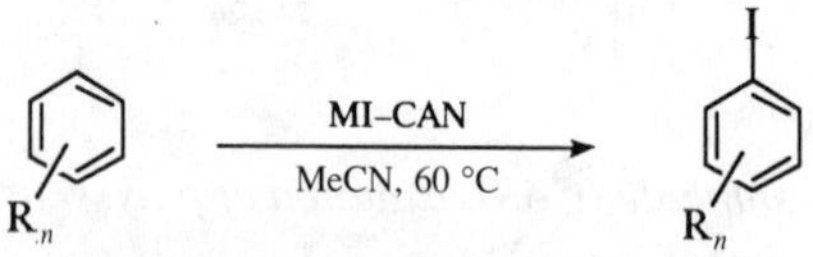

R = Me (n = 2,3,4,5) or OMe (n = 1,2,3)
M = Bu$_4$N, Li, Na

Cerium(IV)-promoted iodination of an aromatic compound. General procedure [7]

To a solution of 4.0 mmol of an aromatic compound and 4.0 mmol of
tetrabutylammonium iodide in 40 ml of acetonitrile was added a solution of 8.0 mmol
of CAN (dried at 90–100 °C for 2 h) in 40 ml of acetonitrile. The mixture was stirred
for 24 h at 60 °C, poured into 80 ml of water, and extracted with benzene (4 x 30 ml).
The combined extracts were washed with sodium thiosulphate solution and water,
and dried over CaCl$_2$. The solvent was carefully removed under reduced pressure and
the residue purified by distillation or chromatography.

Cerium(IV) compounds are effective for the oxidation of side-chains of aromatic
compounds [9–12]. In aqueous medium, the methylene group at the benzylic
position is converted to a carbonyl group. Synthetically useful yields and
selectivities are obtained in most cases. A few examples are shown on the next page.

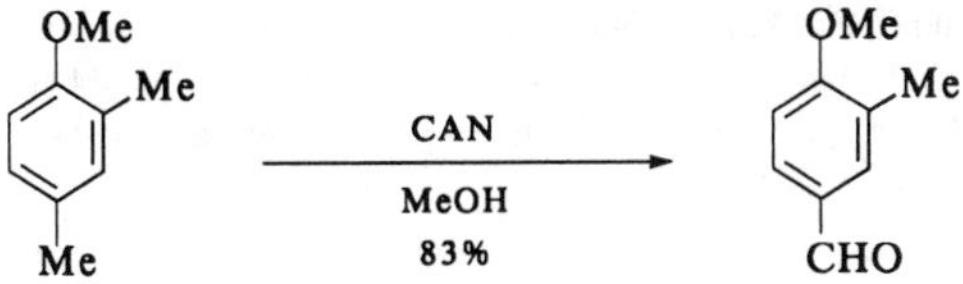

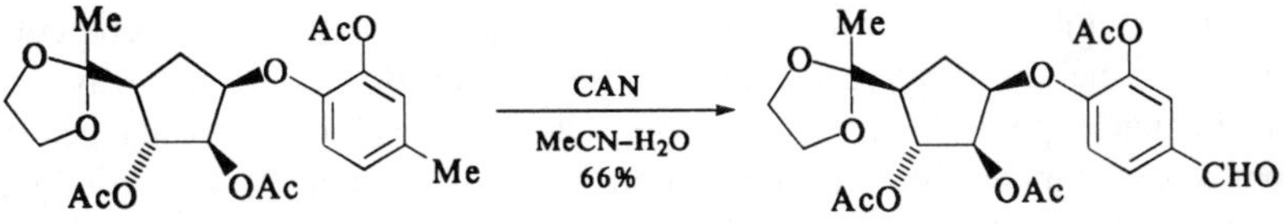

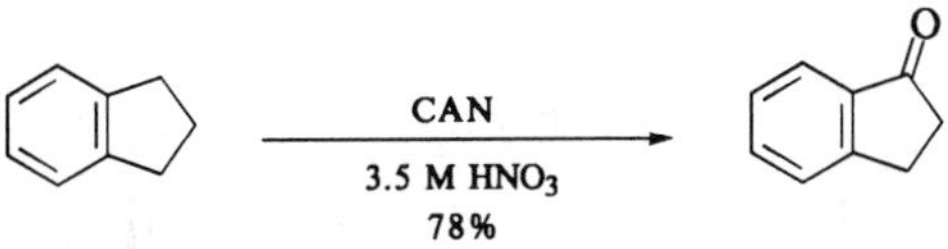

Indirect electrooxidation of alkylbenzenes to aromatic aldehydes and ketones is performed by the recycled use of $Ce(OSO_2CH_3)_3$ or CAN. Generated Ce^{3+} after oxidation of the substrate is efficiently converted to Ce^{4+} by electrooxidation and it is again used for oxidation [4,13].

Oxidation of alkylbenzenes in non-aqueous medium gives products other than carbonyl compounds. Acetates or ethers are produced when the oxidations are carried out in glacial acetic acid or in alcohol [14,15]. Nitrates are obtained in high yields when the substrates and CAN are photolysed in acetonitrile [16].

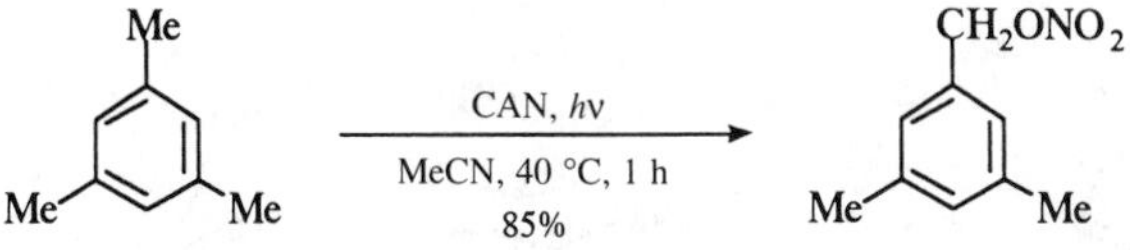

REFERENCES

1.   T.-L. Ho, T.-W. Hall and C. M. Wong, *Synthesis* 206 (1973).
2.   M. Periasamy and M. V. Bhatt, *Synthesis* 330 (1976).
3.   G. Balanikas, N. Hussain, S. Amin and S. S. Hecht, *J. Org. Chem.* **53**, 1007 (1988).
4.   R. P. Kreh, R. M. Spotnitz and J. T. Lundquist, *J. Org. Chem.* **54**, 1526 (1989).
5.   H. M. Chawla and R. S. Mittal, *Synthesis* 70 (1985).
6.   T. Sugiyama, *Chem. Lett.* 1013 (1987).
7.   T. Sugiyama, *Bull. Chem. Soc. Jpn* **54**, 2847 (1981).
8.   C. Galli, *J. Org. Chem.* **56**, 3238 (1991).
9.   S. B. Laing and P. J. Sykes, *J. Chem. Soc. [C]* 2915 (1968).
10.   L. Syper, *Tetrahedron Lett.* . 4493 (1966).
11.   N. Chida, M. Ohtsuka, K. Nakazawa and S. Ogawa, *J. Org. Chem.* **56**, 2976 (1991).
12.   L. K. Sydnes, S. H. Hansen, I. C. Burkow and L. J. Saethre, *Tetrahedron* **41**, 5205 (1985).
13.   T. Imamoto, Y. Koide and S. Hiyama, *Chem. Lett.* **1990**, 1445.
14.   S. Torii, H. Tanaka, T. Inokuchi, S. Nakane, M. Akada, N. Saito and T. Sirakawa, *J. Org. Chem.* **47**, 1647 (1982).

15. E. Baciocchi and C. Rol, *Gazz. Chim. Ital.* **113,** 727 (1983).
16. A. Della Cort, A. L. Barbera and L. Mandolini, *J. Chem. Res. (S)* 44 (1983).
17. E. Baciocchi, C. Rol, G. V. Sebastiani and B. Serena, *Tetrahedron Lett.* **25,** 1945 (1984).

6.2.2 Olefins

Olefins are readily subjected to one-electron oxidation with cerium(IV) compounds, the products being largely dependent on the cerium(IV) compound, solvent and coexisting nucleophile [1–5]. Some examples including a photochemical reaction are shown below. Among these, azidonitration [6], halogenation [7], alkoxyiodination [8] and nitratoiodination [9] are synthetically useful.

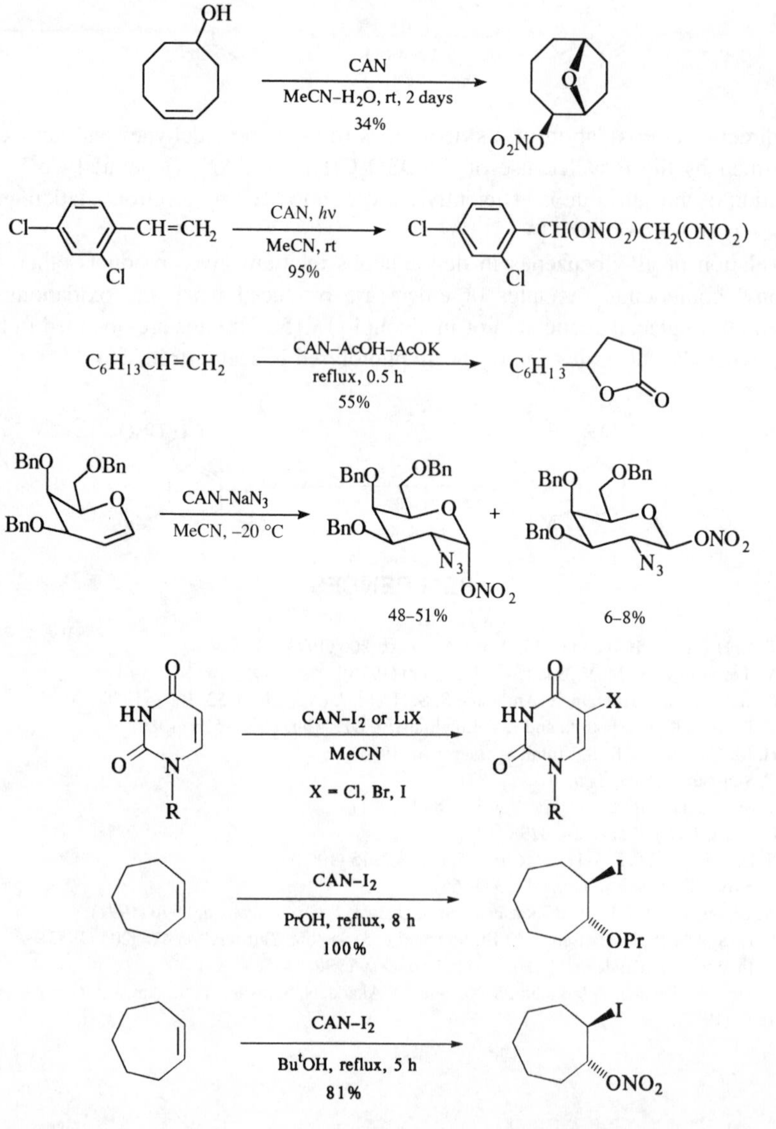

*Preparation of 2-azido-2-deoxy-1,3,4,6-tetra-O-acetyl-β-L-glucopyranose and
2-azido-2-deoxy-1,3,4,6-tetra-O-acetyl-β-l-mannopyranose [6]*

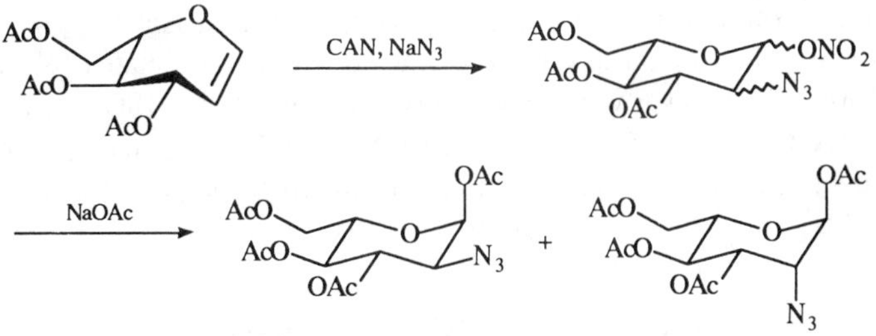

A solution of tri-*O*-acetyl-D-glucal (1.50 g, 5.5 mmol) in acetonitrile (30 ml) was added dropwise to a solid mixture of NaN$_3$ (0.54 g, 8.3 mmol) and CAN (9.1 g, 16.5 mmol) at about –20 °C. The suspension was stirred vigorously under nitrogen. After 7 h, the starting material disappeared on TLC, and the solution was poured into ice water and extracted with ethyl acetate. The combined extracts were successively washed with water and brine, dried over anhydrous MgSO$_4$ and concentrated *in vacuo*. The compound was chromatographed on a flash silica gel column with ethyl acetate–toluene (1 : 12) to give 1.57 g of the azido compound. The ^{1}H NMR spectrum showed at least three isomers (gluco and manno types which contained their α and β isomers).

The azido compound (1.37 g, 3.6 mmol) prepared above was dissolved in acetic acid (30 ml), and sodium acetate (0.60 g, 7.3 mmol) was added to this solution. The mixture was heated at 100 °C overnight. After cooling, the reaction mixture was diluted with ethyl acetate (30 ml), then successively washed with water, aqueous NaHCO$_3$ and brine, dried over anhydrous MgSO$_4$ and concentrated. The residue was chromatographed on a flash column with ethyl acetate–toluene (1 : 5) to give two products. The less polar isomer was identified as the gluco-type compound, 2-azido-2-deoxy-1,3,4,6-tetra-*O*-acetyl-β-L-glucopyranose (0.78 g), and the more polar isomer was identified as the manno-type compound, 2-azido-2-deoxy-1,3,4,6-tetra-*O*-acetyl-β-L-mannopyranose (0.75 g).

Iodination of acetylated uracil nucleoside [7]

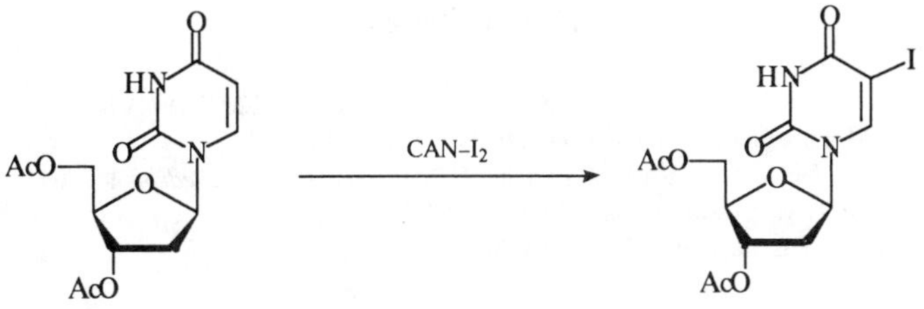

(1)　　　　　　　　　　　　　　(2)

A mixture of acetylated uracil nucleoside (compound **1**) (0.156 g, 0.5 mmol), iodine (0.076 g, 0.3 mmol), CAN (0.137 g, 0.25 mmol) and MeCN (8 ml) was stirred at 80 °C for 1 h. Reaction progress was monitored by TLC ($CHCl_3$–acetone, 4 : 1). After completion of the reaction, the solvent was evaporated and the residue partitioned between a cold mixture of ethyl acetate (20 ml), saturated NaCl (10 ml), and 5% $NaHSO_3$ (5 ml). The aqueous layer was extracted with ethyl acetate (2 x 10 ml). The combined organic layer was washed carefully with cold 5% $NaHSO_3$ (5 ml) followed by saturated NaCl (15 ml) and water (2 x 15 ml), dried ($MgSO_4$) and evaporated. The crude product was recrystallized from EtOH to give compound **2** as colourless needles (0.201 g, 92%), m.p. 157–159 °C.

Reaction of cyclohexene with iodine and CAN in methanol [8]

A mixture of cyclohexene (0.500 g, 6.09 mmol), iodine (0.382 g, 1.51 mmol), CAN (0.849 g, 1.55 mmol) and methanol (50 ml) was stirred at 50 °C for 6 h. The mixture was concentrated and poured into water. The product was extracted with ether and the extract washed with water, dried and evaporated. The residue was purified by distillation to give *trans*-1-iodo-2-methoxycyclohexane (0.331 g, 91% based on iodine), b.p. 97–98 °C/11 mmHg.

Reaction of cyclohexene with iodine and CAN in t-butanol [8]

A mixture of cyclohexene (0.300 g, 1.38 mmol), iodine (0.465 g, 1.83 mmol), CAN (1.00 g, 1.82 mmol) and t-butanol (30 ml) was stirred at reflux for 5.5 h. After the usual work-up, the resulting oil was purified by distillation to give *trans*-2-iodocyclohexanol nitrate (0.792 g, 80% based on cyclohexene), b.p. 125–127 °C/ 12 mmHg.

REFERENCES

1.   C. Briguet, C. Freppel, J.-C. Richer and M. Zador, *Can. J. Chem.* **52**, 3201 (1974).
2.   H. Kim and M. F. Schlecht, *Tetrahedron Lett.* **29**, 1771 (1988).
3.   E. Baciocchi, T. Del Giacco, S. M. Murgia and G. V. Sebastiani, *Tetrahedron* **44**, 6651 (1988).
4.   E. I. Heiba and R. M. Dessau, *J. Am. Chem. Soc.* **93**, 995 (1971).
5.   K. Briner and A. Vasella, *Helv. Chim. Acta* **70**, 1341 (1987).

6.   C.-H. Lin, T. Sugai, R. L. Halcomb, Y. Ichikawa and C.-H. Wong, *J. Am. Chem. Soc.* **114**, 10138 (1992).

7.   J. Asakura and M. J. Robins, *J. Org. Chem.* **55**, 4928 (1990).

8.   C. A. Horiuchi, Y. Nishio, D. Gong, T. Fujisaki and S. Kiji, *Chem. Lett.* 607 (1991).

6.3   OXIDATIONS OF OXYGEN FUNCTIONALITIES

6.3.1   Alcohols

A large variety of oxidation patterns are observed in the cerium(IV) oxidation of alcohols. Various examples are tabulated in Table 6.1.

Benzylic and allylic alcohols are most easily oxidized under mild conditions to carbonyl substrates. Acyloins are also readily converted to α-diketones. More rigorous conditions are required for the oxidation of simple secondary alcohols. Simple primary alcohols resist oxidation by Ce(IV) compounds. Oxidations of tertiary alcohols accompany C–C bond fission to yield fragmentation products. Vicinal diols also undergo oxidative cleavage under standard conditions.

Interestingly, selective oxidation of secondary alcohols in the presence of primary ones is accomplished by the use of catalytic cerium salts. Sodium bromate, oxygen and t-butyl hydroperoxide are all utilized as stoichiometric oxidants in these transformations.

Oxidation of benzylic and related alcohols by aqueous CAN. General procedure [1]

Into a 250 ml Erlenmeyer flask equipped with a magnetic stirring bar was placed 19 mmol of the alcohol and 40 ml of water. If the alcohol was a liquid at room temperature or was water-soluble, 40 ml of a 1 M solution of CAN was added with stirring at room temperature. If the alcohol was a water-insoluble solid at room temperature, the alcohol–water mixture was heated and stirred until the solid melted and then the CAN solution was added with stirring. The mixture or solution was heated and stirred at temperatures up to 90 °C to complete the reaction. The initial colour of the reaction mixture was a deep red which faded to yellow or colourless when reaction was complete. (There are some exceptions to this colour change, however.) When reaction was complete, the reaction mixture was cooled and extracted three times with ether or dichloromethane. The combined extracts were washed with saturated NaHCO$_3$ solution and dried over MgSO$_4$. The solvent was removed by distillation on a steam bath through a 20 cm column packed with glass helices. The last traces of solvent were removed on a rotary evaporator. The purity of the product was determined by NMR. In some cases the product was purified by distillation.

TABLE 6.1

Oxidation of Alcohols by Cerium(IV) Compounds

Alcohol	Reagent and conditions	Product(s) (%)	Ref.
PhCH$_2$OH	**CAN** AcOH–H$_2$O (1 : 1), 90 °C, 8 min	PhCHO (94)	[1]
(pyrimidinedione bearing CH$_2$OH and Me)	**CAN** H$_2$O, 25 °C	(pyrimidinedione bearing CHO and Me) (60)	[2]
(cyclopentane-1,3-diol with CH$_2$CH$_2$OH side chain)	CAN (10 mol %)–NaBrO$_3$ MeCN–H$_2$O, 80 °C, 0.5 h	(hydroxyethyl cyclopentanone) (89)	[3]
(carveol)	CAN/NAFK (2.7 mol %)–Bu^{t}OOH C$_6$H$_6$, 80 °C, 6 h	(carvone) (98)	[4]
(bicyclo[2.2.2]octanol)	**CAN** MeCN–H$_2$O, 50 °C, 0.5 h	CH$_2$CHO (46) + CH$_2$CHO–ONO (29) + CH$_2$CHO–ONO$_2$ (20)	[5]

Substrate	Reagents	Product	Ref.
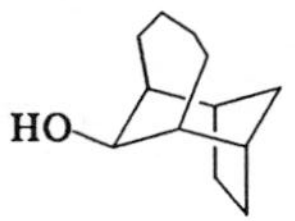 HO– (bicyclic alcohol)	CAN MeCN–H$_2$O, 60 °C	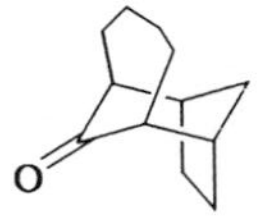 (bicyclic ketone) (90)	[6]
 HO– (bicyclic alcohol)	CAN MeCN–H$_2$O, 60 °C	 (bicyclic ether) (85)	[6]
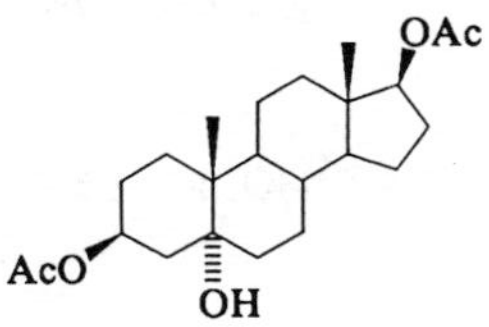 OAc / AcO (steroidal diol)	CAN MeCN–H$_2$O, 80 °C	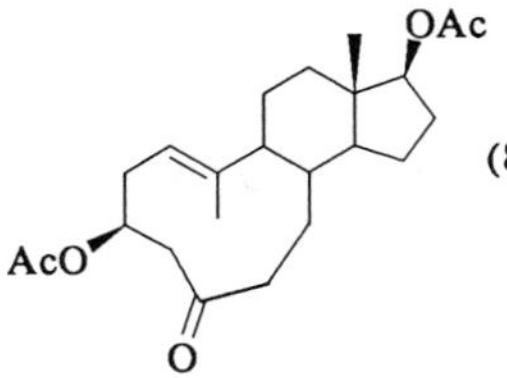 OAc / AcO (ring-expanded ketone) (80)	[7]
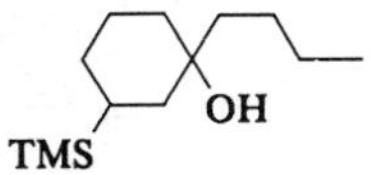 TMS / OH (cyclohexanol)	CAN MeCN–H$_2$O, rt	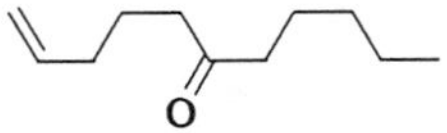 O (ketone) (66)	[8]
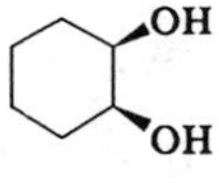 OH / OH (cyclohexanediol)	Ce(ClO$_4$)$_4$ H$_2$O	OHC(CH$_2$)$_4$CHO (98.5)	[9]

TABLE 6.1 *Continued*

Alcohol	Reagent and conditions	Product(s) (%)	Ref.
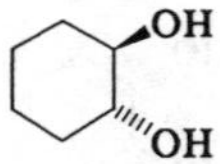 (cyclohexane-1,2-diol, OH, OH)	Ce(ClO$_4$)$_4$ H$_2$O	OHC(CH$_2$)$_4$CHO (91.3)	[9]
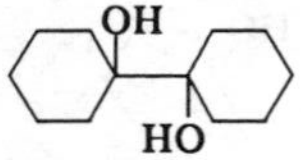 (1,1'-bicyclohexyl-1,1'-diol, OH, HO)	CAN AcOH–H$_2$O, 25 °C, 2h	2 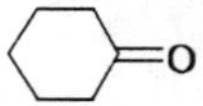 (cyclohexanone) (94)	[10]
PhCHOHCOPh	CAN MeCN–H$_2$O, 60 °C, 10 min	PhCHO (80) + PhCO$_2$H (86)	[11]
PhCHOHCOPh	CAN (15 mol %)–charcoal, air toluene, 100 °C, 2 h	PhCOCOPh (86)	[12]

Oxidation of benzylic and related alcohols by CAN in 50% aqueous acetic acid.
General procedure [1]

Into a 250 ml Erlenmeyer flask equipped with a magnetic stirring bar was placed a solution of 10 mmol of the alcohol in 40 ml of glacial acetic acid. To this was added 40 ml of a 1 M solution of CAN in water. The resulting solution or mixture was heated to temperatures of up to 90 °C to complete the reaction. The initial colour of the reaction mixture was deep red and faded to yellow or colourless at the completion of the oxidation. (There are exceptions to this colour change, however.) The reaction mixture was cooled, diluted with two volumes of water, and extracted three times with ether or dichloromethane. The combined extracts were washed with water and 1.5 M KOH or saturated NaHCO$_3$. The organic layer was dried over MgSO$_4$ and the solvent removed by distillation on a steam bath through a 20 cm column packed with glass helices. The last traces of solvent were removed on a rotary evaporator. The purity of the product was determined by NMR. In some cases the product was purified by distillation.

CAN-catalysed oxidation of 3-(2-hydroxyethyl)cyclopentanol with NaBrO$_3$ [3]

To a suspension of NaBrO$_3$ (0.27 g, 1.8 mmol) in aqueous acetonitrile were added 3-(2-hydroxyethyl)cyclopentanol (0.23 g, 1.8 mmol) and CAN (0.1 g, 0.18 mmol). The mixture was heated at reflux for 30 min and the solvent was removed under reduced pressure. The residue was diluted with ethyl acetate and poured into water. After extraction with ethyl acetate, the combined organic layer was dried over Na$_2$SO$_4$. Concentration followed by purification by silica gel column chromatography (hexane–ethyl acetate, 3 : 1) gave 3-(2-hydroxyethyl)cyclopentanone (0.20 g, 89%).

Oxidation of carveol with t-butyl hydroperoxide using Ce(IV) – NAFK as catalyst
[4]

(i) Preparation of Ce(IV)–NAFK   To a solution of CAN (1.75 g, 3 mmol) in water (30 ml) was added NAFK (1.1 g, 1 mmol of sulphonate), and the whole was stirred for 36 h at 25 °C. Titration of the dried resin indicated that 0.54 mmol of Ce(IV) per gram was loaded on the resin catalyst.

(ii) Oxidation of carveol   Carveol (0.15 g, 1 mmol) and t-butyl hydroperoxide (2.64 M benzene solution, 1.5 ml, 4 mmol) were added to a suspension of Ce(IV)–NAFK (0.050 g, 0.027 mmol of cerium(IV) salt) in benzene (2 ml). The mixture was stirred at 80 °C for 2 h. The resulting mixture was diluted with ethyl acetate, then Ce(IV)–NAFK catalyst was filtered off and washed with aqueous NaHSO$_3$ and brine, and dried over Na$_2$SO$_4$. Concentration gave a residual oil which was subjected to chromatography on silica gel to give carvone (0.15 g, 98%).

Air oxidation of 4-chlorobenzyl alcohol catalysed by CAN–charcoal [12]

(i) *Preparation of the catalyst*   Activated charcoal (3.0 g) was added to a solution of CAN (0.82 g, 1.5 mmol) in water (100 ml) with stirring. After 10 min, the solid was collected by filtration and dried over P_2O_5 *in vacuo* for 15 h to give 3.8 g of catalyst.

(ii) *Oxidation of 4-chlorobenzyl alcohol*   A mixutre of 4-chlorobenzyl alcohol (0.143 g, 1 mmol), the catalyst (0.36 g) and toluene (3 ml) was heated with stirring at 100 °C in an open flask. After 2 h, the reaction mixture was cooled and the catalyst was removed by filtration. The filtrate was concentrated and the residue purified by preparative TLC on silica gel to give 4-chlorobenzaldehyde (0.117 g, 82%).

REFERENCES

1.   W. S. Trahanovsky, L. B. Young and G. L. Brown, *J. Org. Chem.* **32**, 3865 (1967).
2.   D.-R. Hwang, P. Helquist and M. S. Shekhani, *J. Org. Chem.* **50**, 1264 (1985).
3.   S. Kanemoto, H. Tomioka, K. Oshima and H. Nozaki, *Bull. Chem. Soc. Jpn* **59**, 105 (1986).
4.   S. Kanemoto, H. Saimoto, K. Oshima, K. Utimoto and H. Nozaki, *Bull. Chem. Soc. Jpn* **62**, 519 (1989).
5.   W. S. Trahanovsky, P. J. Flash and L. M. Smith, *J. Am. Chem. Soc.* **91**, 5068 (1969).
6.   Y. Fujise, E. Kobayashi, H. Tsuchida and S. Ito, *Hetrocycles* **11**, 351 (1978).
7.   V. Balasubramanian and C. H. Robinson, *Tetrahedron Lett.* **22**, 501 (1981).
8.   S. R. Wilson, P. A. Zucker, C. Kim and C. A. Villa, *Tetrahedron Lett.* **26**, 1969 (1985).
9.   H. L. Hintz and D. C. Johnson, *J. Org. Chem.* **32**, 556 (1967).
10. W. S. Trahanovsky, L. H. Young and M. H. Bierman, *J. Org. Chem.* **34**, 869 (1969).
11. T.-L. Ho, *Synthesis* 560 (1972).
12. Y. Hatanaka, T. Imamoto and M. Yokoyama, *Tetrahedron Lett.* **24**, 2399 (1983).

6.3.2   Phenols and phenol ethers

6.3.2.1   Oxidation of hydroquinones

Hydroquinones are most easily oxidized by CAN. However, the reactions often suffer from overoxidation by the strong oxidizing ability of CAN, to give complex reaction mixtures. Modified procedures employing silica gel-coated CAN [1] or a dual oxidant system CAN–NaBrO3 [2] are successfully employed for the preparation of quinones.

Oxidation of hydroquinone by cerium(IV)–SiO$_2$ reagent [1]

(i) Preparation of cerium(IV)–SiO$_2$ reagent To an efficiently stirred solution of CAN (20 g) in methanol (200 ml), dichloromethane (200 ml) followed by silica gel (40–140 mesh, 200 g) was added. The mixture was stirred for an additional 15 min and the solvents were evaporated on a rotary evaporator. The resulting solid was dried for 3 h at 55 °C on the rotary evaporator, when a free-flowing yellow powder was obtained. At this stage the yellow powder of cerium(IV)–SiO$_2$ reagent weighed 221.2 g, suggesting that 1.2 g of the solvent was retained. For this reagent, 6.06 g contain 1 mmol of cerium(IV).

(ii) Oxidation of hydroquinone by cerium(IV)–SiO$_2$ reagent Hydroquinone (0.110 g, 1 mmol) in dichloromethane (15 ml) was added with stirring to the cerium(IV)–SiO$_2$ reagent (6.94 g, 2.1 mmol of cerium(IV)). The mixture was stirred for 5 min and filtered. The residue was washed with dichloromethane (2 x 15 ml). Evaporation of solvent from the combined filtrates gave 1,4-benzoquinone (0.106 g, 98%) as a bright-yellow crystalline solid. The product was further purified by crystallization from ether, m.p. 116 °C.

REFERENCES

1. A. Fischer and G. N. Henderson, *Synthesis* 641 (1985).
2. T.-L. Ho, *Synth. Commun.* **9**, 237 (1979).

6.3.2.2 Oxidation of phenol ethers

Cerium(IV) ammonium nitrate (CAN) is capable of oxidizing phenol ethers to quinones [1]. 4-Dimethoxybenzene derivatives are most easily oxidized to 4-benzoquinone derivatives in good yields. In polynuclear aromatic systems, the most electron-rich ring system is preferentially oxidized, as exemplified by the following scheme [1–4]. Pyridinedicarboxylic acid *N*-oxide acts as an effective ligand, providing high yields of quinone products [3,5,6].

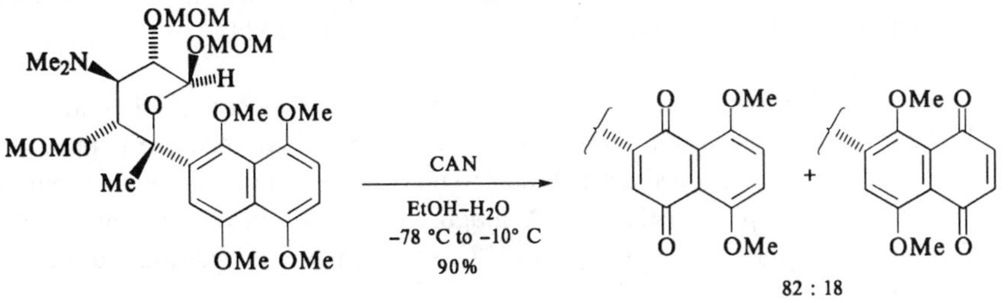

Preparation of 1,4-benzoquinone from 1,4-dimethoxybenzene [1]

To a solution of 1,4-dimethoxybenzene (1.38 g, 10 mmol) in acetonitrile (25 ml) was added a solution of CAN (16.5 g, 30 ml) in water (25 ml). The mixture was extracted with chloroform. The extract was concentrated under reduced pressure, and the residue was sublimed (160 °C/20 mmHg) to give the yellow 1,4-quinone (0.61 g, 57%), m.p. 111.5–112.5 °C.

Oxidation of compound 3 by CAN in the presence of pyridinedicarboxylic acid N-oxide [3]

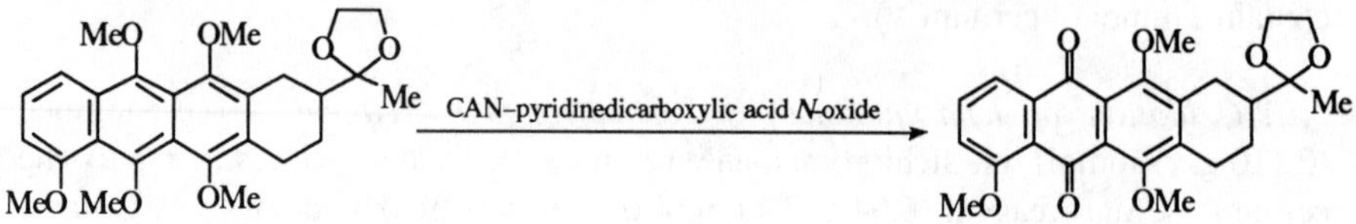

(3)   (4)

(i) Preparation of pyridinedicarboxylic acid N-oxide [6] Pyridine-2,6-dicarboxylic acid (16.7 g, 100 mmol) and a solution of $Na_2WO_4 \cdot 2H_2O$ (1.0 g, 3.3 mmol) in 30% hydrogen peroxide (50 ml) were heated at 90–100 °C with vigorous stirring for 50 min. Then more 30% hydrogen peroxide (120 ml) was added in portions over a period of 2 h until all insoluble substance vanished. After an additional 3 h of heating, the reaction mixture was allowed to stand for several hours. The crystalline solid was filtered, washed with several portions of the cold water and dried in the air. The filtrate was extracted with chloroform. Evaporation of the solvent gave a residue which was combined with the main portion of crude N-oxide. The pure pyridine-2,6-dicarboxylic acid N-oxide (13.8 g, 75%) was obtained by recrystallization of the crude product from water, m.p. 158–160 °C.

(ii) Oxidation of compound 3 [3] To a magnetically stirred, cold (0 °C) solution of compound 3 (5.20 g, 11 mmol) in a mixture of acetonitrile (150 ml) and THF (150 ml) were sequentially added powdered pyridinedicarboxylic acid N-oxide (5.05 g, 27 mmol) and a solution of CAN (15.22 g, 27.8 mmol) in water (35 ml). The reaction mixture was stirred for 10 min, and then partitioned between ethyl acetate (250 ml) and brine (200 ml). The organic layer was washed with water (2 x 50 ml), dried (MgSO₄) and filtered, and the solvent was evaporated to give the crude quinone as a yellow solid. Recrystallization of the crude product from ether–hexane gave pure quinone 4 (4.68 g, 96%) as long, yellow needles, m.p. 147–152 °C.

Quinone formation reactions using CAN proceed under almost neutral conditions, and other functional groups are tolerated in most cases. This procedure is therefore valuable for the total synthesis of natural products possessing a quinone system. A typical example is depicted below [7]. Analogous procedures have been successfully applied to the synthesis of fridamycin E [8], sarubicin A [9] and quinoid alcohol dehydrogenase coenzyme [10,11].

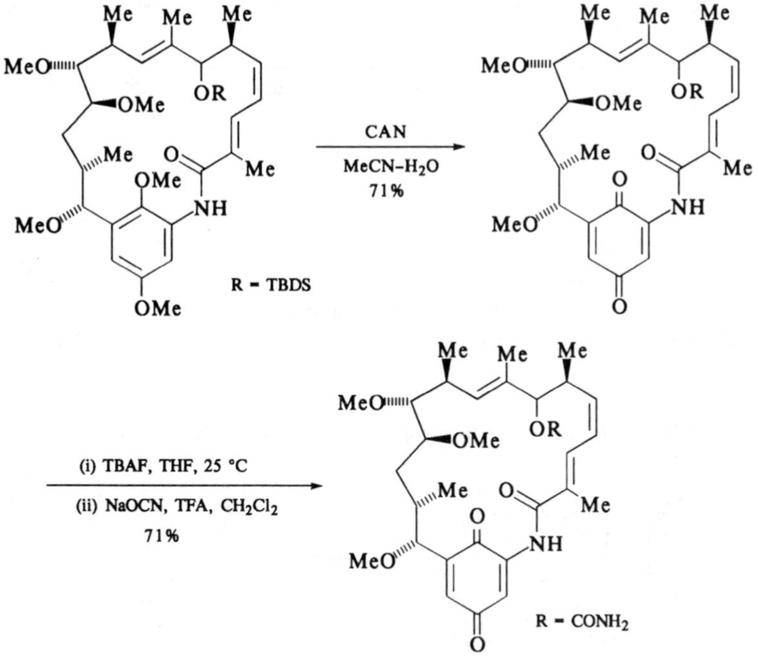

CAN

MeCN–H₂O

71%

R = TBDS

(i) TBAF, THF, 25 °C

(ii) NaOCN, TFA, CH₂Cl₂

71%

R = CONH₂

Macbecin I

REFERENCES

1. P. Jacob III, P. S. Callery, A. T. Shulgin and N. Castagnoli Jr, *J. Org. Chem.* **41**, 3627 (1976).
2. Y. Tanoue and A. Terada, *Bull. Chem. Soc. Jpn* **61**, 2039 (1988).
3. F. M. Hauser and S. Prasanna, *J. Am. Chem. Soc.* **103**, 6378 (1981).
4. M. Kawasaki, F. Matsuda and S. Terashima, *Tetrahedron* **44**, 5713 (1988).
5. R. P. Thummel, S. Chirayil, C. Hery, J.-L. Lim and T.-L. Wang, *J. Org. Chem.* **58**, 1666 (1993).
6. L. Syper, K. Kloc, J. Mlochowski and Z. Szulc, *Synthesis* 521 (1979).
7. D. A. Evans, S. J. Miller, M. D. Ennis and P. L. Ornstein, *J. Org. Chem.* **57**, 1067 (1992).
8. T. Matsumoto, H. Jona, M. Katsuki and K. Suzuki, *Tetrahedron Lett.* **32**, 5103 (1991).
9. Y. Takeuchi, M. Sudani and E. Yoshii, *J. Org. Chem.* **48**, 4151 (1983).
10. E. J. Corey and A. Tramontano, *J. Am. Chem. Soc.* **103**, 5599 (1981).
11. S. Itoh, J. Kato, T. Inoue, Y. Kitamura, M. Komatsu and Y. Ohshiro, *Synthesis* 1067 (1987).

6.3.2.3 Oxidative removal of phenol ether protecting groups

The characteristic oxidizing ability of CAN is utilized for the removal of protecting groups. 4-Methoxyphenyl, 2,4-dimethoxyphenyl, 4-methoxybenzyl and 2,4-dimethoxybenzyl groups are readily removed by treatment with CAN at around room temperature. The oxidative removal occurs chemoselectively and hence this protocol can be utilized in diverse synthesis. Representative examples are shown below [1,2].

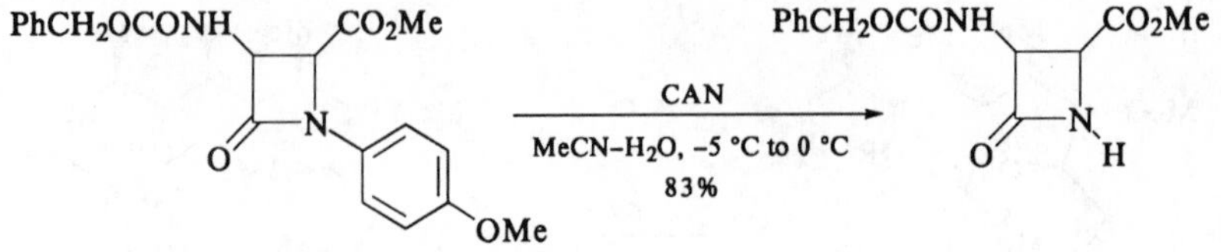

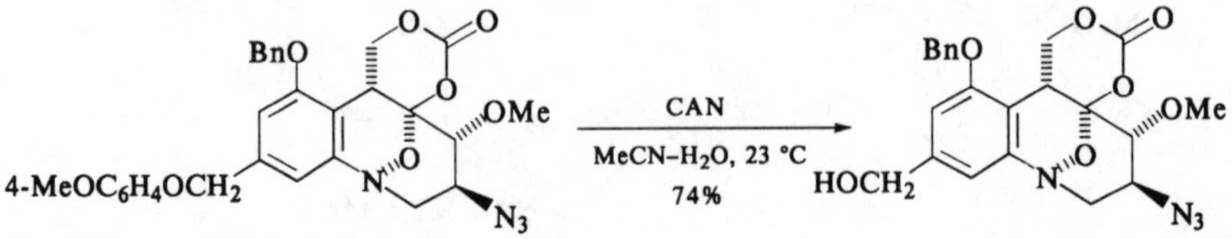

Reaction of cis-*3-azido-1-(4-methoxyphenyl)-2-oxo-4-(2-phenylethenyl)-1-azetidine with CAN [1]*

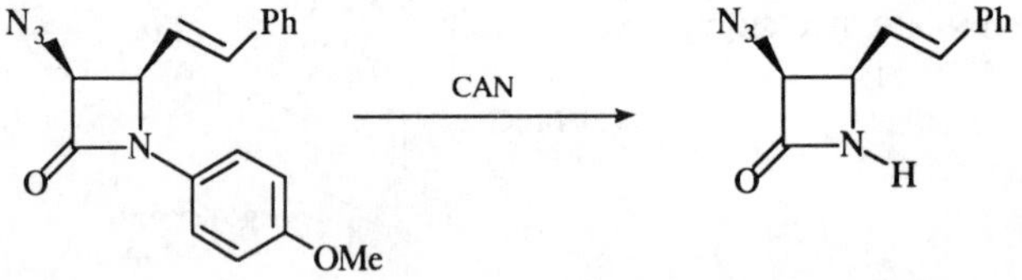

A solution of *cis*-3-azido-1-(4-methoxyphenyl)-2-oxo-4-(2-phenylethenyl)-1-azetidine (4.62 g, 14.4 mmol) in acetonitrile (150 ml) was cooled to 0 °C and treated with a solution of CAN (23.7 g, 43.3 mmol) in water (200 ml) for 3 min. The reaction was stirred at –5 °C to 0 °C for 25 min and diluted with 1 litre of water. The mixture was extracted with ethyl acetate (3 x 200 ml). The orange extracts were washed with 5% NaHCO₃ (500 ml) and the aqueous extracts were back-washed with ethyl acetate (100 ml). The combined organic solutions were washed with 10% sodium sulphite, 5% NaHCO₃ (100 ml) and brine. The resulting solution was swirled over charcoal for 30 min, sodium sulphate was added, and the mixture was filtered through Celite. Removal of the solvent yielded a yellow oil which was dissolved in ether and cooled. Filtration yielded *cis*-3-azido-2-oxo-4-(2-phenylethenyl)-1-azetidine as 1.91 g of white solid. And additional 0.23 g of material was obtained from the mother liquor. The total yield was 68%, m.p. 103–105 °C.

REFERENCES

1.   D. R. Kronenthal, C. Y. Han and M. K. Taylor, *J. Org. Chem.* **47**, 2765 (1982).
2.   T. Fukuyama, L. Xu and S. Goto, *J. Am. Chem. Soc.* **114**, 383 (1992).

6.3.3   Enol silyl ethers and related substrates

Oxidation of enol silyl ethers with CAN affords 1,4-diketones [1]. Regioselective coupling between dienol silyl ethers and enol silyl ethers to provide 6-oxo-α,β-unsaturated carbonyl compounds is also promoted by CAN, as exemplified by the following scheme [2].

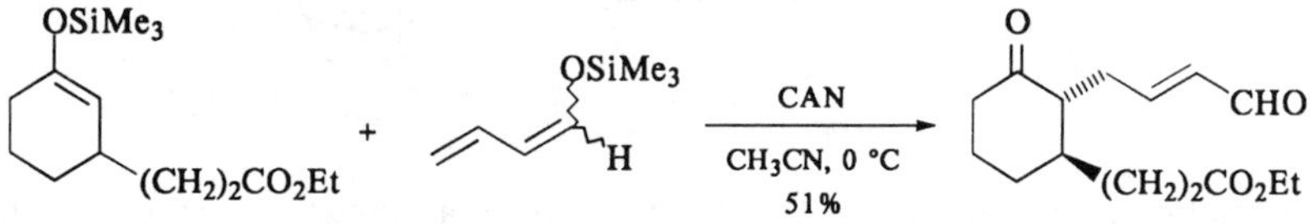

This CAN-promoted coupling reaction is applicable for the synthesis of 4-oxoaldehyde from enol trimethylsilyl ethers and ethyl vinyl ether [3].

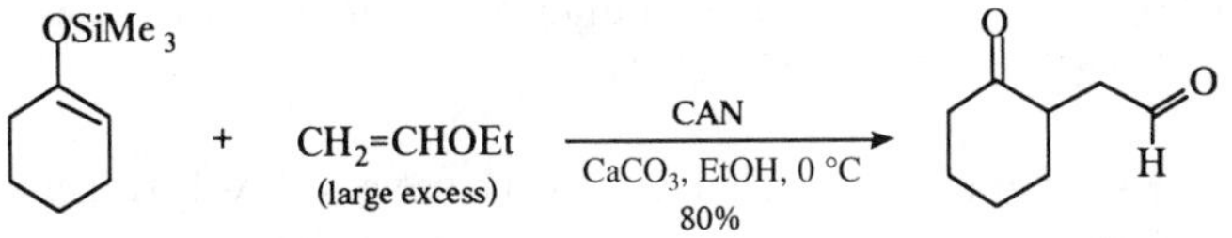

Enol silyl ethers of unsaturated aromatic ketones are subjected to oxidative cyclization by CAN, as exemplified by the following scheme [4]. The yields of the products, however, are somewhat lower than those obtained by the use of Cu(OTf)$_2$.

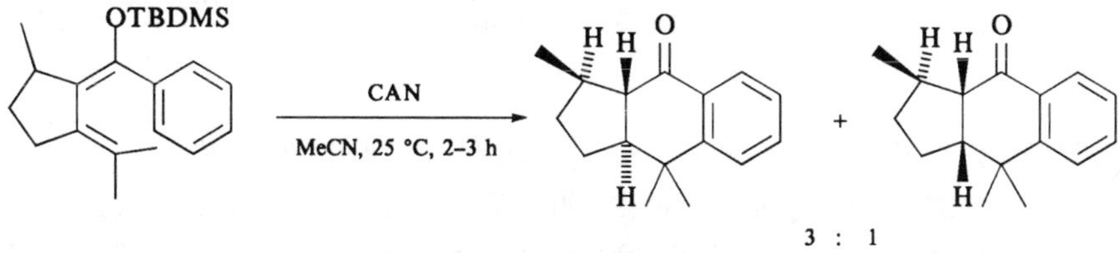

3 : 1

REFERENCES

1. E. Baciocchi, A. Casu and R. Ruzziconi, *Tetrahedron Lett.* **30**, 3707 (1989).
2. A. B. Paolobelli, D. Latini and R. Ruzziconi, *Tetrahedron Lett.* **34**, 721 (1993).
3. E. Baciocchi, A. Casu and R. Ruzziconi, *Synlett* 679 (1990).
4. B. B. Snider and T. Kwon, *J. Org. Chem.* **57**, 2399 (1992).

6.3.4 Carbonyl compounds and related substrates

Oxidation of ketones with CAN proceed through C–C bond cleavage. Simple aliphatic ketones are converted to a mixture of nitrato carboxylic acids [1]. Yields and selectivity are not satisfactory in most cases. On the other hand, some cyclic ketones, particularly polycyclic ketones, undergo Baeyer–Villiger type oxidation with CAN to give lactones in good yields [1–3]. In contrast to the classical Baeyer–Villiger reaction, this CAN-promoted reaction is completed within a short time.

Oxidation of adamantanone by CAN [1]

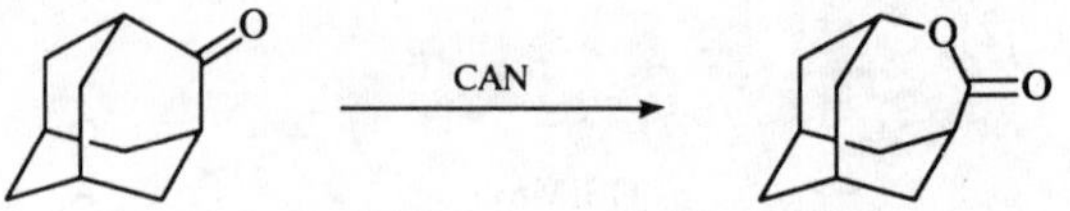

(5)

To a solution of adamantanone (1.0 g, 6.6 mmol) in acetonitrile (15 ml) was added CAN (15.0 g, 27.3 mmol) in water (30 ml). The resulting solution was stirred magnetically at 60 °C for 3 h during which time the orange colour faded to a faint yellow. The solution was cooled, poured into brine and extracted with chloroform. The extracts were dried and evaporated to give a soft, solid mass. Sublimation of this material at 130–150 °C (bath temperature) under water aspirator pressure furnished lactone 5 (0.815 g, 73%), m.p. 285–287 °C.

Direct α-iodination of ketones is achieved by the use of CAN–I_2 reagent system [4]. In the case of unsymmetrical ketones, the more substituted position is preferentially iodinated.

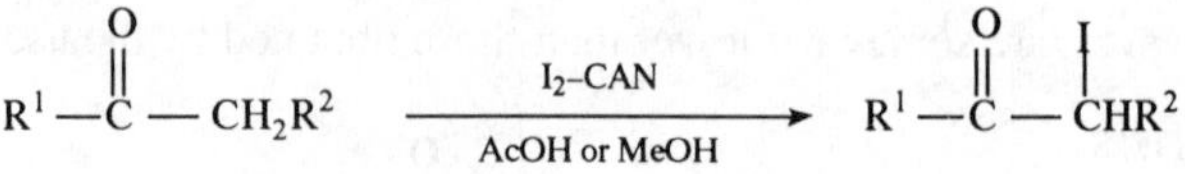

Preparation of 2α-iodo-5α-cholestan-3-one from 5α-cholestan-3-one [4]

A mixture of 5α-cholestan-3-one (0.200 g, 0.517 mmol), iodine (0.066 g, 0.259 mmol) and CAN (0.142 g, 0.259 mmol) in acetic acid–water (20 ml, 9 : 1) was stirred at 50 °C for 8 h. The reaction mixture was poured into water and extracted with ether. The ethereal solution was washed with aqueous $NaHCO_3$ and water, dried and concentrated. Crystallization of the residue from ethanol gave needles of 2α-iodo-5α-cholestan-3-one (0.213 g, 80%), m.p. 132–133 °C.

The reaction of cyclohexenone derivatives with CAN–I_2 in alcohol provides alkyl phenyl ethers in good yields [5]. In the case of isophorone, an oxidative 1,2-rearrangement of the C_5–methyl group occurs, providing 5-alkoxy-1,2,3-trimethylbenzene derivatives.

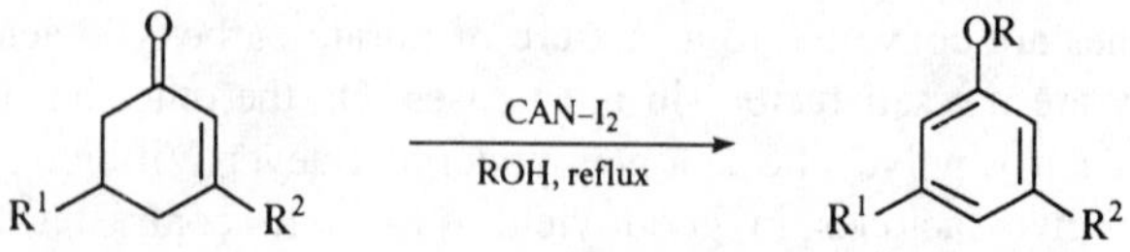

Simple alkanoic acids are resistant to Ce(IV) oxidation. However, if an α-substituent that stabilizes a free radical is present, the carboxylic acid undergoes facile decarboxylation. For example, α-hydroxycarboxylic acids are readily subjected to oxidative decarboxylation by Ce(IV) salts to give ketones [6–8].

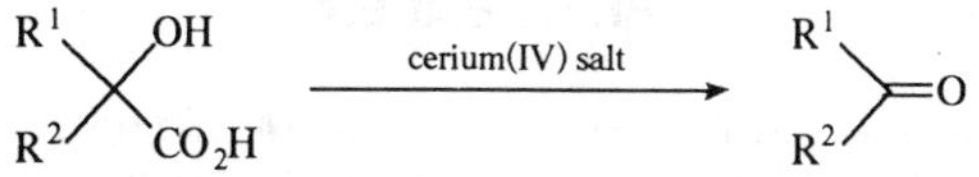

Similarly, α-alkoxymalonic and α-hydroxymalonic acids undergo oxidative bisdecarboxylation on treatment with CAN [9,10]. This reaction, in combination with the pericyclic reaction of olefins with diethyl oxomalonate, provides a useful entry for the synthesis of lactones.

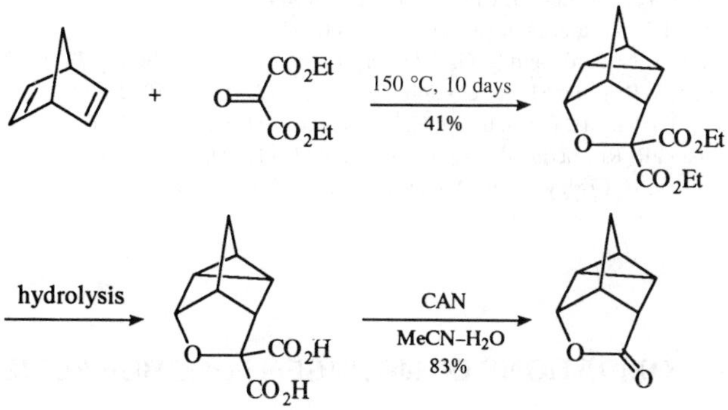

Dialkyl malonates react with cerium(IV) salts to generate malonyl radicals, which are trapped by alkenes or aromatic compounds [11,12]. These reactions are utilized for the introduction of malonyl components into heterocyclic compounds [13].

Preparation of dimethyl (2-thienyl)propanedioate [13]

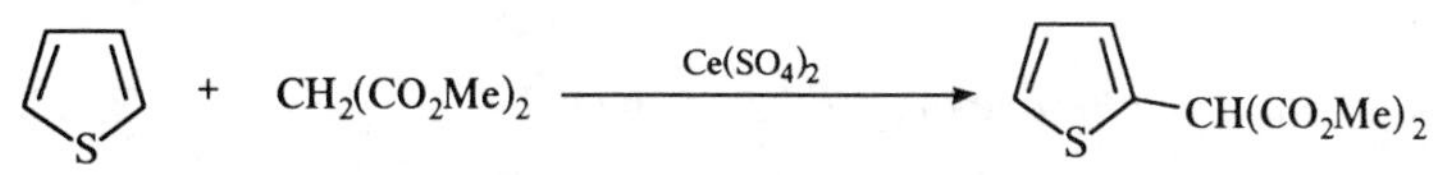

A solution of thiophene (2 ml, 25 mmol) and dimethyl malonate (5 ml, 45 mmol) in 50 ml of methanol–water (9 : 1) was treated with cerium (IV) sulphate (4.04 g, 10 mmol). The heterogeneous mixture was stirred at ambient temperature under nitrogen until a negative starch iodine test was obtained (4 h). The solids were then filtered from the reaction mixture and washed with methanol (25 ml). The combined filtrate was concentrated to 10 ml at reduced pressure and then partitioned between dichloromethane (50 ml) and water (50 ml). The organic layer was washed with water (50 ml), then with saturated brine (50 ml). After drying over Na_2SO_4, the solvent was removed *in vacuo*. The excess dimethyl malonate was distilled off at 60 °C/0.5 mmHg. Column chromatography of the residue on silica gel (50 g) using dichloromethane as the eluent gave dimethyl (2-thienyl)propanedioate (1.82 g, 85%) as a colourless oil.

REFERENCES

1. P. Soucy, T.-L. Ho and P. Deslongchamps, *Can. J. Chem.* **50**, 2047 (1972).
2. G. Mehta, P. N. Pandey and T.-L. Ho, *J. Org. Chem.* **41**, 953 (1976).
3. T.-L. Ho, T. W. Hall and C. M. Wong, *Synth. Commun.* **3**, 79 (1973).
4. C. A. Horiuchi and S. Kiji, *Chem. Lett.* 31 (1988).
5. C. A. Horiuchi, H. Fukunishi, M. Kajima, A. Yamaguchi, H. Kiyomiya and S. Kiji, *Chem. Lett.* 1921 (1991).
6. B. Krishna and K. C. Tewari, *J. Chem. Soc.* 3097 (1961).
7. T. J. Kemp and W. A. Waters, *J. Chem. Soc.* 1192 (1964).
8. S. B. Hanna and S. A. Sarac, *J. Org. Chem.* **42**, 2069 (1977).
9. M. F. Salomon, S. N. Pardo and R. G. Salomon, *J. Am. Chem. Soc.* **102**, 2473 (1980).
10. R. G. Salomon, S. Roy and M. F. Salomon, *Tetrahedron Lett.* **29**, 769 (1988).
11. E. Baciocchi, B. Giese, H. Farshchi and R. Ruzziconi, *J. Org. Chem.* **55**, 5688 (1990).
12. E. Baciocchi and R. Ruzziconi, *J. Org. Chem.* **56**, 4772 (1991).
13. L. M. Weinstock, E. Corley, N. L. Abramson, A. O. King and S. Karady, *Heterocycles* **27**, 2627 (1988).

6.4   OXIDATIONS OF NITROGEN FUNCTIONALITIES

Primary or secondary nitro compounds can be transformed to aldehydes or ketones in high yields when treated with CAN in the presence of triethylamine [1,2].

$$R^1 \diagdown \atop R^2 \diagup CH-NO_2 \xrightarrow[\text{Et}_3\text{N, H}_2\text{O}]{\text{CAN}} R^1 \diagdown \atop R^2 \diagup {=}O$$

Oxidation of nitrocyclohexane to cyclohexanone [1]

To a well-stirred mixture of nitrocyclohexane (0.65 g, 5 mmol) and triethylamine (5 ml) in acetonitrile (14 ml) in a 50 ml round-bottomed flask fitted with a water condenser was added an aqueous solution of CAN (2.75 g, 5 mmol) in water (6 ml). The solution turned into a deep-brown emulsion, which was then heated to 50 °C and maintained at that temperature until completion of the reaction. The progress of the reaction was monitored by taking periodic aliquots and analysing the ethereal extracts after work-up by infra-red spectroscopy. After completion of the reaction, the mixture was cooled, diluted with acetonitrile (20 ml) and filtered to remove all insoluble materials. The filtrate was taken up in ether (2 × 50 ml) and washed successively with water, diluted hydrochloric acid and brine. The dried extracts were evaporated to give cyclohexanone (0.40 g, 80%), b.p. 46–48 °C/15 mmHg.

Oximes, semicarbazones and carboxylic acid hydrazides are rapidly oxidized by CAN to give parent carbonyl compounds or carboxylic acids [3,4]. Carboxylic acid amides derived from 5,6-dihydrophenanthridine are readily oxidized by CAN to produce the parent carboxylic acids in excellent yields [5].

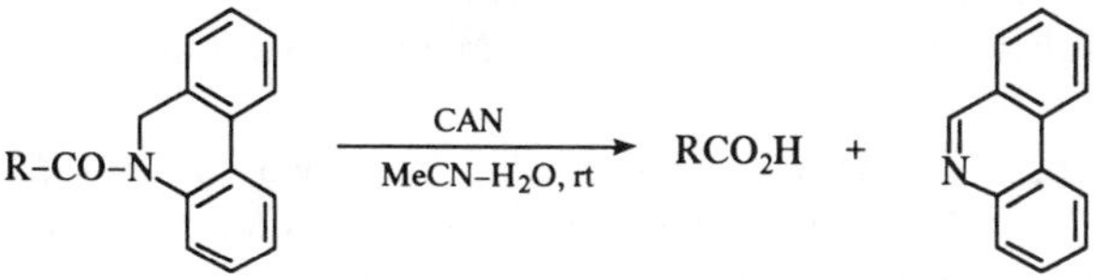

Oxidation of 5-(4-phenylbutanoyl)-5,6-dihydrophenanthridine [5]

To a solution of 5-(4-phenylbutanoyl)-5,6-dihydrophenanthridine (0.245 g, 0.45 mmol) in acetonitrile–water (7.5 ml, 4 : 1) was added CAN (0.904 g, 1.65 mmol) all at once. After being stirred for 15 min, the reaction mixture was diluted with ether. The organic layer was washed successively with 1 M aqueous HCl solution and brine, and dried over Na_2SO_4. The solvent was removed under reduced pressure to give pure 4-phenylbutanoic acid (0.118 g, 97%).

Enamines react with Ce(IV) salts to generate nitrogen cation radicals. These reactive intermediates are trapped by electron-rich olefins to afford addition products. Use of tetrabutylammonium cerium(IV) nitrate, $(Bu_4N)_2Ce(NO_3)_6$, in place of CAN gives higher yields of the products. An example is shown below [6].

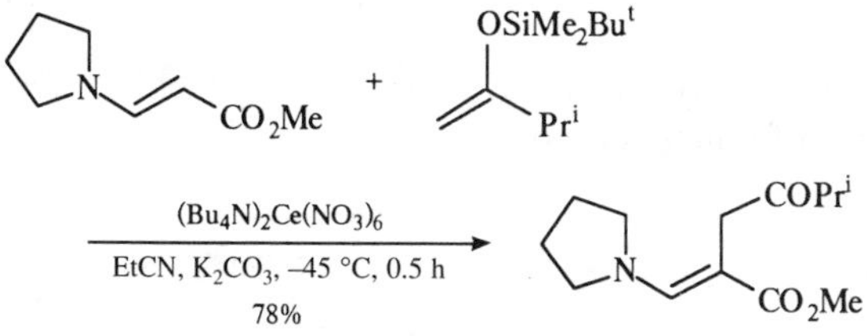

REFERENCES

1.   G. A. Olah and B. G. B. Gupta, *Synthesis,* 44 (1980).
2.   R. C. Cookson and P. S. Ray, *Tetrahedron Lett.* **23**, 3521 (1982).
3.   J. W. Bird and D. G. M. Diaper, *Can. J. Chem.* **47**, 145 (1969).
4.   T.-L. Ho, H. C. Ho and C. M. Wong, *Synthesis* 562 (1972).
5.   T. Uchimaru, K. Narasaka and T. Mukaiyama, *Chem. Lett.* 1551 (1981).
6.   K. Narasaka, T. Okauchi, K. Tanaka and M. Murakami, *Chem. Lett.* 2099 (1992).

6.5   OXIDATIONS OF SULPHUR FUNCTIONALITIES

Thiols are selectively oxidized to disulphides by $[Ce(NO_3)_2]CrO_4$ [1,2]. Aryl sulphides are readily oxidized by CAN to sulphoxides; no sulphones are produced even when excess oxidant is employed [3]. This method, however, is not suitable for oxidation of dialkyl sulphides which possess α-hydrogen atoms, probably due to the susceptibility of the resultant sulphoxides to Pummerer-type rearrangement. Phase transfer conditions using stoichiometric amounts of CAN are quite efficient; various sulphides including dialkyl sulphides are converted to sulphoxides in essentially quantitative yields [4]. Procedures employing catalytic CAN with stoichiometric amounts of $NaBrO_3$ [5] or O_2 [6] are also applicable to the oxidations of dialkyl sulphides.

1,3-Dithiolanes and dithianes are readily degraded to their parent carbonyl compounds [7–8]. The utility of this method is demonstrated by its successful application to the compounds possessing keto olefins [9] or sensitive acetals [10].

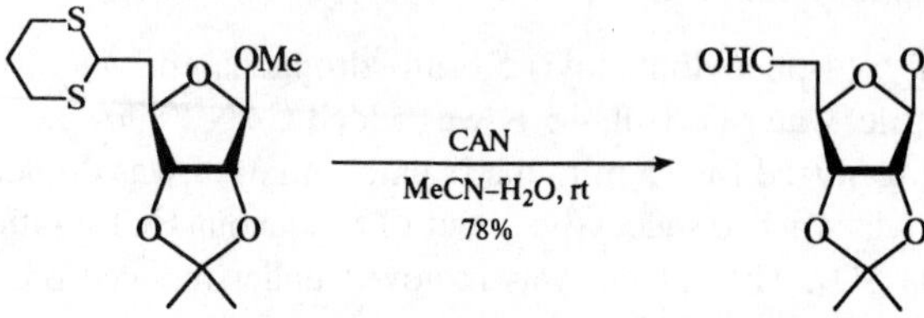

Allyl sulphides react with silyl enol ethers, siloxy dienes or siloxy enynes in the presence of CAN to give α-phenylthio γ,δ-unsaturated ketones [11].

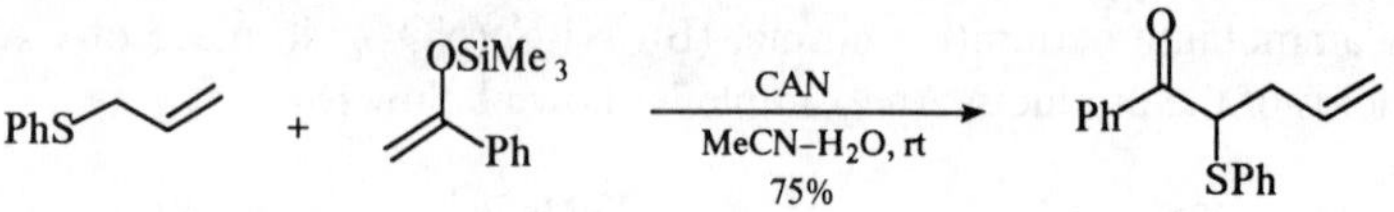

The following reaction pathway was proposed by the authors for this interesting carbon–carbon bond-forming reaction.

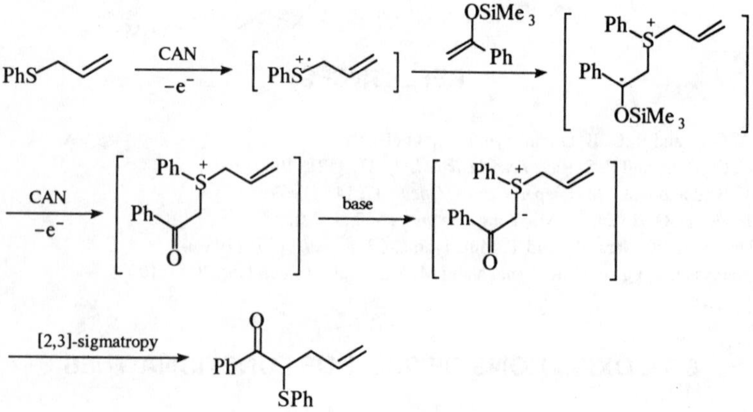

General procedure for dethioacetalization with CAN [7,12]

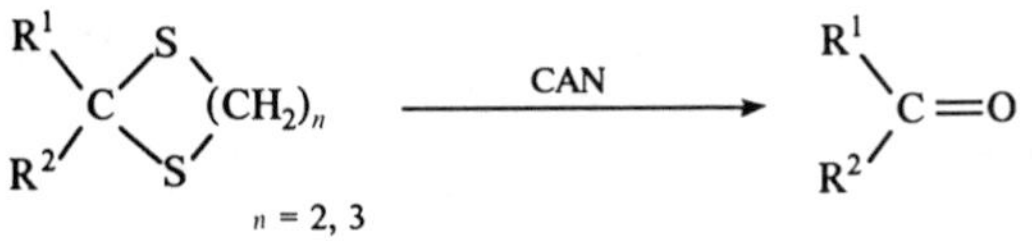

$$\begin{array}{c} R^1 \\ \diagdown \\ C \\ \diagup \\ R^2 \end{array} \begin{array}{c} S \\ \diagdown \\ (CH_2)_n \\ \diagup \\ S \end{array} \xrightarrow{\text{CAN}} \begin{array}{c} R^1 \\ \diagdown \\ C=O \\ \diagup \\ R^2 \end{array}$$

n = 2, 3

A dithioacetal (1 mmol) was dissolved in 75% aqueous acetonitrile and was treated with CAN (4 mmol) at room temperature for 3 min. Addition of water and subsequent extraction with ether gave the carbonyl compound. Pure product was obtained by distillation or by filtration through a neutral alumina column and recrystallization.

REFERENCES

1. H. Firouzabadi and N. Iranpoor, *Synth. Commun.* **14**, 875 (1984).
2. H. Firouzabadi, N. Iranpoor, F. Kiaeezadeh and J. Toofan, *Synth. Commun.* **14**, 973 (1984).
3. T.-L. Ho and C. M. Wong, *Synthesis* 561 (1972).
4. E. Baciocchi, A. Piermattei and R. Ruzziconi, *Synth. Commun.* **18**, 2167 (1988).
5. T.-L. Ho, *Synth. Commun.* **9**, 237 (1979).
6. D. P. Riley, M. R. Smith and P. E. Correa, *J. Am. Chem. Soc.* **110**, 177 (1988).
7. T.-L. Ho, H. C. Ho and C. M. Wong, *J. Chem. Soc. Chem. Commun.* 791 (1972).
8. H. C. Ho, T.-L. Ho and C. M. Wong, *Can J. Chem.* **50**, 2718 (1972).
9. A. M. Sepulchre, G. Vass and S. D. Gero, *Tetrahedron Lett.* 3619 (1973).
10. H. J. Cristau, B. Chabaud and H. Christol, *J. Org. Chem.* **49**, 2023 (1984).
11. K. Narasaka and T. Okauchi, *Chem. Lett.* 515 (1991).
12. T.-L. Ho, *Synthesis* 347 (1973).

6.6 OXIDATIONS OF OTHER HETEROATOM FUNCTIONALITIES

Organosilicon and organotin compounds are subjected to one-electron oxidation by CAN [1–4]. Examples of regiospecific and chemoselective oxidations are shown below.

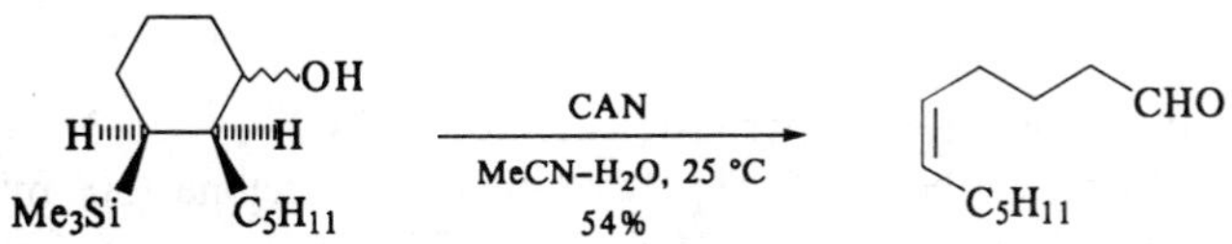

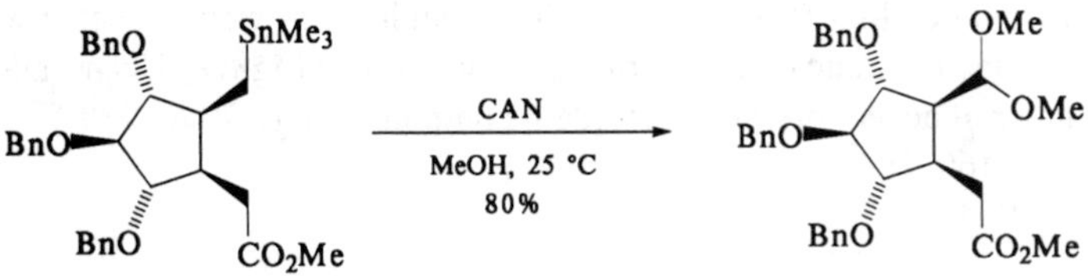

REFERENCES

1.   S. R. Wilson, P. A. Zucker, C. Kim and C. A. Villa, *Tetrahedron Lett.* **26**, 1969 (1985).
2.   S. R. Wilson and C. E. Augelli-Szafran, *Tetrahedron* **44**, 3983 (1988).
3.   S. Hanessian and R. Léger, *J. Am. Chem. Soc.* **114**, 3115 (1992).
4.   S. Hanessian and R. Léger, *Synlett* 402 (1992).

6.7 OXIDATIONS OF METAL COMPLEXES

In 1965 the cerium(IV) oxidation procedure for the liberation of π-ligands from organometallic compounds was employed in the successful generation of cyclobutadiene from its iron tricarbonyl complex [1]. This highlighted reaction has promoted studies on analogous reactions, and a number of decomplexations of various organometallics have been reported so far [2]. Representative examples are tabulated in Table 6.2.

Transition metal complexes of simple olefins undergo facile decomplexation to yield parent olefins. Arene complexes of chromium also undergo decomplexation by CAN to aromatic substrates. When the molecule possesses hydroquinone moieties, further oxidation proceeds to give a quinone. σ-Bonded organometallics of iron, cobalt, manganese, molybdenum and tungsten are readily oxidized to carboxylic acid derivatives. Fischer-type carbene complexes are also subjected to CAN oxidation to give esters or lactones. Acyl complexes of iron are converted to lactones or lactams. The procedure is a useful entry for the synthesis of β-lactones and β-lactams.

Reaction of cyclobutadieneiron tricarbonyl with CAN in the presence of 1,4-thiapyron-1,1-dioxide [3]

(6)

To a solution of 1,4-thiapyron-1,1-dioxide (2.00 g, 15.3 mmol) and cyclobutadieneiron tricarbonyl (2.92 g, 15.3 mmol) in acetone (500 ml) at room temperature was added powdered CAN (30 g, 55 mmol) over a 4 min period. Vigorous stirring was continued for an additional minute, after which the reaction mixture was poured into 500 ml of brine. The solution was well extracted with ether and the combined organic extracts dried over anhydrous $MgSO_4$. Evaporation of the solvent under reduced pressure afforded compound **6** (2.70 g, 98%) as yellow crystals, m.p. 160–165 °C.

TABLE 6.2

Oxidation of Transition Metal Complexes with Cerium(iv) Compounds

Substrate	Reagent and conditions	Product(s) (%)	Ref.
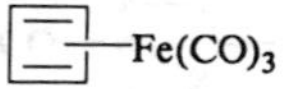 $Fe(CO)_3$	CAN acetone, −5 °C to rt	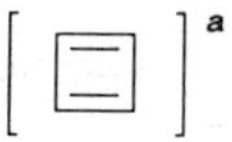 [a]	[1–4]
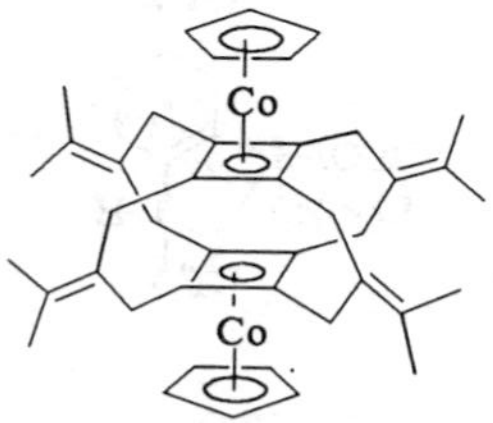	CAN cyclohexane–MeCN, rt, 18 h	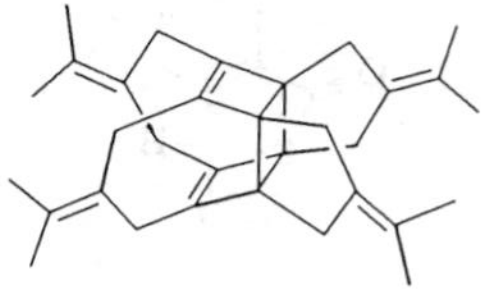 (88)	[5]
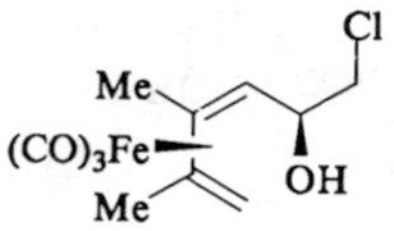 $(CO)_3Fe$, Me, Me, Cl, OH	CAN EtOH–H_2O, −10 °C	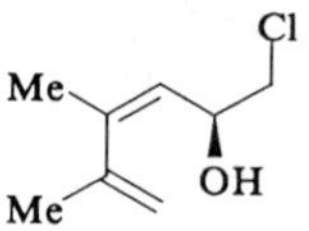 Cl, OH, Me, Me (97)	[6]

Continued

TABLE 6.2 *Continued*

Substrate	Reagent and conditions	Product(s) (%)	Ref.
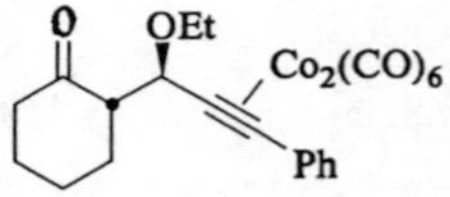 (OEt, Co₂(CO)₆, Ph, cyclohexanone structure)	CAN–Et₃N acetone, 0 °C	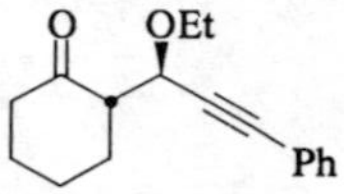 (84)	[7]
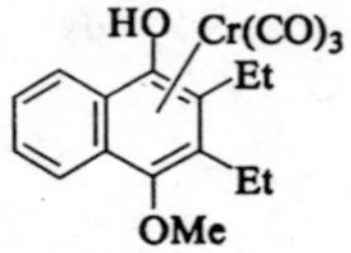 (HO, Cr(CO)₃, Et, Et, OMe naphthalene structure)	CAN HNO₃–H₂O, 25 °C, 30 min	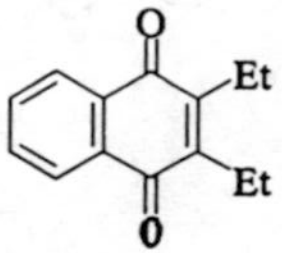 (96)	[8]
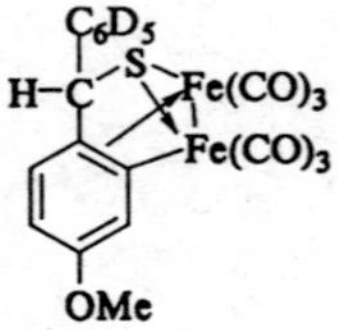 (C₆D₅, S, Fe(CO)₃, Fe(CO)₃, OMe structure)	CAN acetone, rt, 2 h	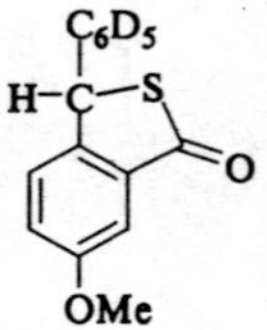 (81)	[9]
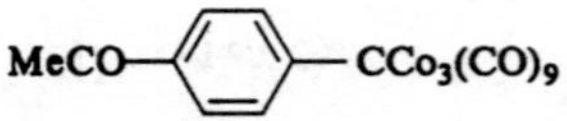 (MeCO, CCo₃(CO)₉ benzene structure)	CAN EtOH, –15 °C	 (88)	[10]
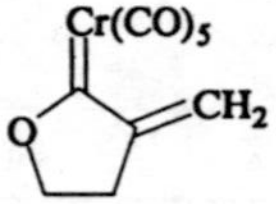 (Cr(CO)₅, CH₂, lactone structure)	CAN, acetone, rt, 1 min	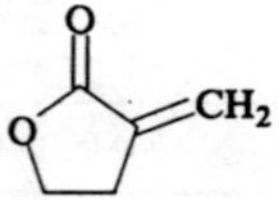 (76)	[11]

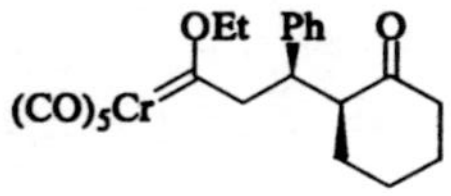

(CO)$_5$Cr

OEt Ph

CAN
acetone, rt

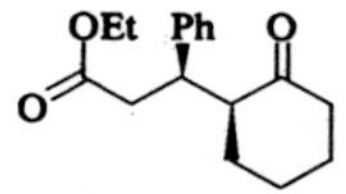

OEt Ph
O

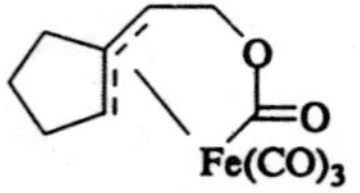

(68)

[12]

Fe(CO)$_3$
O
O

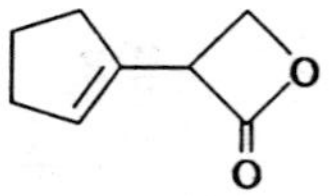

CAN
EtOH, −15 °C

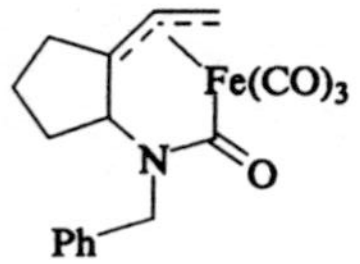

O
O

(100)

[13]

Fe(CO)$_3$
O
N
Ph

CAN
EtOH, −15 °C

O
N
Ph

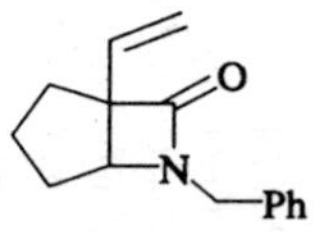

(75)

[14]

<hr>

[a] Cyclobutadine is rapidly dimerized at room temperature.

REFERENCES

1. L. Watts, J. D. Fitzpatrick and R. Pettit, *J. Am. Chem. Soc.* **87,** 3253 (1965).
2. T.-L. Ho, in *Organic Syntheses by Oxidation with Metal Compounds,* ed. W. J. Mijs and C. R. H. L. de Jonge, p. 569. Plenum, New York, 1986.
3. L. A. Paquette and L. D. Wise, *J. Am. Chem. Soc.* **89,** 6659 (1967).
4. L. A. Paquette, Y. Hanzawa, K. J. McCullough, B. Tagle, W. Swenson and J. Clardy, *J. Am. Chem. Soc.* **103,** 2262 (1981).
5. R. Gleiter, R. Merger and B. Nuber, *J. Am. Chem. Soc.* **114,** 8921 (1992).
6. M. Franck-Neumann, A. Abdali, P.-J. Colson and M. Sedrati, *Synlett* 331 (1991).
7. R. Tester, V. Varghese, A. M. Montana, M. Khan and K. M. Nicholas, *J. Org. Chem.* **55,** 186 (1990).
8. W. D. Wulff, P.-C. Tang and J. S. McCallum, *J. Am. Chem. Soc.* **103,** 7677 (1981).
9. H. Alper and A. S. K. Chan, *J. Am. Chem. Soc.* **95,** 4905 (1973).
10. D. Seyferth and A. T. Wehman, *J. Am. Chem. Soc.* **92,** 5520 (1970).
11. C. P. Casey and W. R. Brunsvold, *J. Organometal. Chem.* **102,** 175 (1975).
12. S. Aoki, T. Fujimura and E. Nakamura, *J. Am. Chem. Soc.* **114,** 2985 (1992).
13. G. D. Annis, S. V. Ley, C. R. Self and R. Sivaramakrishnan, *J. Chem. Soc. Perkin Trans. 1,* 270 (1981).
14. G. D. Annis, E. M. Habblethwaite and S. V. Ley, *J. Chem. Soc. Chem. Commun.* 297 (1980).

Index of Compounds and Methods

A

Acetalization, 114
Acetals, 114
Acetone cyanohydrin, 110, 112
Acetonitrile, 116
Acetophenone
  hydrocyanation,111
  Meerwein–Ponndorf–Verley reduction, 75
2-Acetoxy-5-iodo-1-phenyl-1-pentanone, 27
Acetylenic alcohols, 43
Acid chlorides, 43, 54, 89
Acrolein, 104
Acyloins, *see* α-Hydroxy ketones
Acylsamariums, 44
Adamantanone, 138
1,2-Addition
  organocerium reagents, 81, 84,93
Alcohols
  oxidation by Ce (IV) reagents, 127
  preparation, 40-49, 80-97
Aldehydes
  acetalization, 114
  aldol reaction, 98-102
  coupling with alkynes, 56
  [2+2] cycloaddition, 109
  from methylbenzenes, 123
  from nitoalkanes, 140
  from thioacetals, 142, 143
  hetero Diels–Alder reaction, 103-109
  hydrocyanation, 110
  α-hydroxy, 16
  pinacol coupling, 52, 53
  reactions with organocerium reagents, 81, 86
  reductions by lanthanide intermetallics, 18
Aldol reaction, 97, 98-101
Aldols, 76, 97, 98-101
Alkanes
  from organic halides, 24, 78
Alkenes
  from alkynes, 14, 35
  from β-chloro ethers, 24, 26
  hydrogenation, 18, 77
  intramolecular coupling with ketones 56-58
  oxidations by Ce(IV) reagents, 124
α-Alkoxy acid chlorides, 44
Alkoxyiodination, 124
α-Alkoxymalonic acids, 139
Alkyl fluorides, 24, 78
Alkyl halides
  Barbier-type reactions, 40-49
  coupling with isonitriles, 42
  reduction, 7, 24, 25, 78
Alkynes
  coupling with aldehydes or ketones, 56
  reduction to alkenes, 14, 35
Alkynyl alcohols, 88, 89, 93
Allenic alcohols, 43
Alloys (Lanthanide intermetallics), 18
Allylic acetates, 35, 43

Allylic alcohols
  cyclopropanation, 8-12
  oxidation by Ce(IV) reagents, 127
  preparation, 30, 56, 67-71, 79, 86, 87, 93
Allylic samarium (III) complexes, 63
Allylic silanes, 89, 94
2-(Allyloxy)-1-iodobenzene, 59
Allylsamariums (III), 43
Allyl phenyl sulphide, 142
Amides
  α-hydroxy, 16
  reaction with organocerium reagents, 95
  reduction to amines, 33
Amines
  primary, 7, 33, 34, 38, 39, 79
  secondary, 116
Amine N-oxides, 35
Amino acid derivatives, 16
Aminomethyl alcohols, 54
Amino olefins, 115
5-Amino-1-pentene, 116
5-Amino-5-phenylnonane, 83
Anthracene, 121
Aromatic hydrocarbons, 120-123
Arsine oxides, 35
Aryl halides
  reduction, 24, 25, 78
Aryl radicals, 59, 61
Azidonitration, 124
Aziridines, 111
Azobenzene, 64

B

Baeyer–Villiger type oxidation, 137
Barbier-type reaction
  by cerium, 7
  by SmI$_2$, 40-49
Benzaldehydes, 75, 76, 110
  preparation, 123, 127, 128
Benzenes
  polymethyl, 122
Benzhydrol, 75
Benzil, 54
  asymmetric reduction, 33
α-Benzoate esters, 51
Benzofurans, 59
Benzoin
  optically active, 33
Benzonitrile, 83
Benzophenone, 75
Benzoyl chloride, 54
5-Benzoyloxy-2,6-dimethyl-3-heptanol, 76
Benzyl chloromethyl ether, 41
O-Benzyl formaldoxime, 54
Benzylic alcohols, 75, 127, 128
Benzylic samarium (III) complexes, 63
Birch reduction, 14